Happ
Piecewise Methods and Application to Power Systems

Weinberg and Weinberg
On the Design of Stable Systems

On the Design
of Stable Systems

Other Books by Gerald M. Weinberg

1961   *Computer Programming Fundamentals* (with H. D. Leeds) (McGraw-Hill)

1966   *PL/I Programming Primer* (McGraw-Hill)

1970   *Computer Programming Fundamentals Based on the IBM System/360* (with H. D. Leeds) (McGraw-Hill)

1970   *PL/I Programming—A Manual of Style* (McGraw-Hill)

1971   *The Psychology of Computer Programming* (Van Nostrand Reinhold)

1973   *Structured Programming in PL/C* (with N. F. Yasukawa and R. Marcus) (Wiley)

1973   *Teacher's Guide to Structured Programming in PL/C* (with N. F. Yasukawa and R. Marcus)

1975   *An Introduction to General Systems Thinking* (Wiley)

1975   *Structured Programming Film Series* (with Dennis Geller, Naomi Kleid, and Tom Plum) (Edutronics)

1976   *High Level COBOL Programming* (with Steve Wright, Dick Kauffman, and Marty Goetz) (Winthrop)

1977   *Humanized Input—Techniques for Reliable Keyed Input* (with Tom Gilb) (Winthrop)

1978   *EthnoTECHnical Review Handbook* (with Daniel Freedman) (Ethnotech)

1978   *Are Your Lights On?* (with Don Gause) (Ethnotech)

Other Books by Daniela Weinberg

1975   *Peasant Wisdom: Cultural Adaptation in a Swiss Village* (University of California Press)

1975   *Bruson: Etude socio-ethnologique sur les relations humaines dans un village de montagne* (Annales Valaisannes)

On the Design of Stable Systems

A companion volume to
An Introduction to General Systems Thinking
by Gerald M. Weinberg

GERALD M. WEINBERG

Ethnotech, Inc.
Lincoln, Nebraska

DANIELA WEINBERG

University of Nebraska–Lincoln

Illustrated by Sally Cox

A WILEY-INTERSCIENCE PUBLICATION

JOHN WILEY & SONS, New York · Chichester · Brisbane · Toronto

Library of Congress Cataloging in Publication Data

Weinberg, Gerald M
   On the design of stable systems.

   (Wiley series on systems engineering and analysis)
   "A Wiley-Interscience publication."
   Bibliography: p.
   Includes index.
   1. System analysis.   I.   Weinberg, Daniela, joint
author.   II.   Title.

QA402.W43   003   79-13926
ISBN 0-471-04722-8

Printed in the United States of America

10 9 8 7 6 5 4 3 2 1

TO
BEAUTIFUL JOE

SYSTEMS ENGINEERING AND ANALYSIS SERIES

In a society which is producing more people, more materials, more things, and more information than ever before, systems engineering is indispensable in meeting the challenge of complexity. This series of books is an attempt to bring together in a complementary as well as unified fashion the many specialties of the subject, such as modeling and simulation, computing, control, probability and statistics, optimization, reliability, and economics, and to emphasize the interrelationship between them.

The aim is to make the series as comprehensive as possible without dwelling on the myriad details of each specialty and at the same time to provide a broad basic framework on which to build these details. The design of these books will be fundamental in nature to meet the needs of students and engineers and to insure they remain of lasting interest and importance.

Preface

This book is the result of an 18-year collaboration between two people, in two different disciplines, who share a fascination and love for the human animal. Whether from the vantage point of computers or anthropology, we are excited by the capacities of the human mind and alarmed by some of its products.

Our disciplines seem to begin at opposite poles—machine systems and social systems—but they converge as soon as people enter the picture. Thus the social scientist is concerned with the cultural meaning of "math anxiety," while the computer scientist teaches programmers to overcome their "people anxiety."

Both our disciplines daily come to grips with the subtle interplay between system and environment. Cultures and computers both exhibit the effects of adaptation to a constantly changing environment. And anthropologists and computer scientists equally balk at the difficulties of studying conservation and persistence.

This collaboration has been nourished, too, on another level. Our professional activities oscillate between the abstract and the concrete, the theoretical and the practical. The computer scientist designs program logic and enables people to work productively in teams. The anthropologist wrestles with concepts in the classroom and drinks wine with natives in the field. We are accustomed to switching modes, shifting communication styles, all the time. It is perhaps this instability in our professional lives that has taught us to value uncertainty and to make a virtue of indeterminacy. General systems thinking has been, for us, a way of understanding the complexity of our own lives. We sincerely hope it will be the same for you.

How To Use This Book

This book is a companion volume to *An Introduction to General Systems Thinking*, which means that either may be used separately, although they go well together. If you have already read *An Introduction to General Systems Thinking*, you definitely have all the prerequisites you need to read this volume. If you haven't read the companion, you may well find that you will benefit most from reading this volume first, since you have it in your hand.

If you run into any difficulty with the mathematical notation in this volume, try using the Appendix for guidance. If that doesn't solve the problem, and you still feel that the notation is seriously eroding your enjoyment of this volume, then it may be best to set this one aside and read the companion first.

But keep in mind that a major goal of these books is to reduce your anxiety about mathematics, and to open doors that any mathematical deficiency may have closed. If the companion volume is not easily at hand, try reading through this one, ignoring anything you consider too mathematical.

The Questions for Further Research at the end of each chapter should be read as part of the text to give an impression of the scope of problems to which the chapter materials might apply. Should some problem strike you as particularly intriguing, make note of it and then use the references to take it up later. Since the book is intended to introduce you to new ways of thinking, many quotations and references have been given—not to lend a patina of scholarship, but to give you numerous pointers toward other paths to learning.

Following the suggestions of several readers of the earlier volume, we've made a few changes to improve the readability of the book. Chapters have been made shorter and more numerous with more topical divisions to provide places for rest and contemplation. References have been moved to a consolidated bibliography at the end of the book to be less distracting to those reading for the pure pleasure of reading. "Recommended" and "Suggested" readings have been dropped, as have notational exercises, because most people surveyed either ignored them or found them perhaps patronizing. Those readers interested in suggestions for further reading will find numerous clues in the body of the text.

For classroom use, there is a great variety of options. We have used this material in a course of its own, and also in a sequel to a course based on *An Introduction to General Systems Thinking*. Most of the people who use the companion volume as a text have pretty good ideas of their own about how to use it. Many have followed suggestions given in the earlier volume, some of which bear repeating here.

This text and the previous one have been used, to our knowledge, in courses in management science, computer science, anthropology, sociology, urban studies, metallurgy, medicine, architecture, psychology, theology, and philosophy. The texts are suitable for any level, from sophomores on up, if adjustments are made by assigning differing amounts of supplementary reading and questions for further research. The research questions are usually suitable for either a short essay or a term paper. In higher-level courses we have asked students to prepare one or more of these questions for class presentation.

The very flexibility of these books and the generality of their material make them difficult to set in a university curriculum, yet many, many professors have overcome these problems and achieved great success. Some of them have told us that our suggestions in *An Introduction to General Systems Thinking* were useful in getting started. If you are contemplating using this book as a text, look there first. But if you need further advice, the authors would be more than pleased to hear from you and to contribute what they have learned from their own experiences and from contributions of others. In any case, if reading the book itself doesn't give you many ideas, then perhaps we're all in trouble.

Bibliographic Note

The material in this book, accumulated over the better part of two decades, comes from an incredible variety of sources. It has survived more than twenty moves, plus another dozen or so seasonal crossings of the Atlantic. It has been manipulated by more than ten editorial assistants, using at least five bibliographic systems. To top off all this possibility for error, a substantial chunk of the original bibliographic material was misplaced in the most recent move, five years ago.

As a result of these perturbations, the bibliographic material has not been entirely stable. When preparing this manuscript for the printer, we discovered the large loss and a number of smaller problems. In order to reconstruct the bibliography, we and our assistants undertook several hundred hours of detective work, most of which was successful. In the end, however, there remain bibliographic problems which we cannot seem to correct with any reasonable amount of effort.

For example, some of the references are from editions of books that are not available anywhere in Nebraska, and it has proved impossible to check (or even provide) all page numbers. Several references to articles have been almost completely lost, with only the author's name and perhaps a date remaining. Sometimes even that much is missing.

Rather than omit material for which we are unable to provide exact, checked references, we have decided to give what reference material we have and to beg our readers to help us, wherever possible, to improve this information for subsequent printings. We intend no discourtesy to the original authors, and certainly do not offer their original material as ours. We have included direct quotations, even in those few cases where we have lost track of the author. We felt that authors would prefer that their thoughts be spread widely rather than being omitted because of an unfortunate bibliographic accident. We especially hope that we will be able to fill in all the blanks before several printings have passed.

In all, the problems have affected perhaps 2 percent of the bibliographic items, and most of those are merely missing page numbers. We don't think the problem will in any way affect the general usefulness of the book, but we apologize for any inconvenience it may cause any reader.

Acknowledgments

Many people have contributed in many ways to the publication of this book. We wish to express our special thanks to the following friends who have been closest to the project over the years: Sheila Abend, Lorri Campbell, Jim Greenwood, Mike Gunn, Linda Hollcroft, Joan Kaufmann, Shanna McGoff, Cheryl Plum, Mona Thompson, Melissa Weiksnar. Encouraging, questioning, searching out literature, criticizing, typing, preparing the manuscript—in all these ways, their help has been invaluable. We also owe an important debt to our many students, and to the readers of the previous volume who took the time to send us their comments.

GERALD M. WEINBERG
DANIELA WEINBERG

Eagle, Nebraska
May 1979

Contents

On the Design
of Stable Systems

1

The Problem
of Persistence

There is a widespread assumption in modern social science that social continuity requires no explanation. . . .

The assumption of inertia, that cultural and social continuity do not require explanation, obliterates the fact that both have to be recreated anew in each generation, often with great pain and suffering. To maintain and transmit a value system, human beings are punched, bullied, sent to jail, thrown into concentration camps, cajoled, bribed, made into heroes, encouraged to read newspapers, stood up against a wall and shot, and sometimes even taught to read sociology. (Moore, 1966)

Barrington Moore is right about social continuity, but wrong about social science. It's not that social science assumes that social continuity *requires* no explanation. There are deep-seated reasons why social sciences, more so than other sciences, evade or disguise the explanation of continuity.

If Moore is right, we have here a widespread nonpractice—something that continues in many places at many times. This nonpractice itself, then, by Moore's argument, requires an explanation. It does not continue by mere "inertia." It is not explained by the term "assumption." To maintain and transmit this assumption, this nonpractice, there must be activities of some sort—perhaps no heroes or bullies, punching or shooting, but activities all the same.

It will clear the way for the rest of our volume if we devote this first chapter to the question of why continuity is seldom explained, and why we choose to explain it now, in terms of general systems thinking.

Weinberg's Law(s) of Twins

What vanity needs for its satisfaction is glory, and it is easy to have glory without power. The people who enjoy the greatest glory in the United States are film stars, but they can be put in their place by the committee for Un-American Activities, which enjoys no glory whatever.

Bertrand Russell,
Nobel Prize Acceptance Speech

Every systems thinker harbors a secret dream—to be immortalized in the name of some law—Darwin's Principle of Natural Selection, Newton's Law of Universal Gravitation, Shannon's Noisy Channel Theorem, Heisenberg's Uncertainty Principle, Weinberg's Law of Twins.

Weinberg's Law of Twins?! Alas, long before your authors acquired the name "Weinberg," that one slim thread of immortality was forever broken for them. An outstanding mathematician, remembered today for many important contributions to genetics and statistics, stated a principle concerning the frequency of twin births among all births, a principle remembered today as *Weinberg's Law of Twins*.

To any later Weinberg, the existence of this law was an impediment to broad generalization. Knowing that Weinberg's Law was already taken, what was the use of dreaming up another one? Then one day, a realization came. Since the law was Weinberg's Law of Twins, some other worthy law could be given the name of Weinberg's Law, as long as it wasn't a law of twins. After all, there was the Hardy–Weinberg Law, which nobody confused with the other one, the one about twins.

And so the search for the key to fame and fortune resumed with the vigor of a long caged beast. And finally, it came—a law so worthy that it predicted *over 99% of all human behavior*. Surely, of all laws, this one deserved immortality.

Like many truly great laws, this one had the most humble beginnings. The Weinbergs were sitting on the Number 44 bus, heading up Broadway in the gloom of a winter rush hour, when a haggard but pretty young woman boarded with eight children in tow.

"How much is the fare?" she asked the driver.

"Thirty-five cents for adults, and children under five ride free."

"Okay," she said, shifting one of the two tiniest she carried under her arms so she could reach her purse. She dropped two coins into the meter and started with her entourage down the aisle.

"Wait just a minute, lady!" the driver commanded, as only a New York driver can. "You don't expect me to believe that all eight of them children is under five!"

"Of course they are," she said indignantly. "These two are four, the two girls are three, the toddlers are two, and these little ones are one."

The driver was dumbfounded, and apologetic. "Gee lady, I'm sorry. Do you *always* have twins?"

"Oh, no," she said, managing to straighten a wisp of brown hair. "Most of the time we don't have *any*."

And there it was, in a flash, to both of us at once. Most of the time, for most systems in the world, nothing of *any* significance happens. Indeed, if you look one minute and then look the next, most of the time you see almost exactly the same thing. It's what Barrington Moore called "social continuity," but it's far more general than that. For most systems, of any kind, the best prediction you can make for their behavior in the next instant will be that they will be doing just what they were doing the previous instant.

Needless to say, we were elated. Our fame was assured by the discovery of what had to become the most important of all laws, transcending the discoveries of any discipline, of mathematics, of philosophy herself. And then, just as suddenly as the elation had come, gloom descended. For although the new law was certainly *important* enough to carry the name Weinberg's Law, it was, like the other one, a law about the frequency of *twin births!* It could not, therefore, be given its proper name, Weinberg's Law of Twins, for that name was already enscribed in the History of Science.

We were heartbroken, so much so that we didn't tell anybody about our law for many years. When we finally screwed up the courage, our friends merely laughed. "Why don't be silly," they chided us. "There are any number of precedents you can follow. Why, just think of Newton's First and Second Laws of Motion."

And so it was that our chance for ascending to the Valhalla of Science was reborn. There, etched in some marble throne, are *Weinberg's First Law of Twins* and *Weinbergs' Second Law of Twins.* Leaving aside the fancy mathematics, in general systems terms the First Law of Twins says:

Among all births, twin births are pretty rare, and triplets rarer still.

The Second Law—*our* law—says:

Even though twin births are rare among all births, births themselves are almost infinitely rarer than no births at all.

As with Newton's Laws of Motion, the Second Law here, if we must say so ourselves, is by far the more important of the two, in spite of any implications of the word "second." It is important because it does, indeed, *predict* 99% of human behavior, and about the same amount of the behavior of other systems as well. Consider the average woman in the United States. She lives to be 70.7 years and gives birth to 2.1 children. This means that on any day of her life, chosen at random, her chance of giving birth *at all* is less than .01%. (For men, of course, the odds are even lower.) On the other hand, if she does give birth on a certain day, her chance of having twins or better is about 1.5%. Therefore, the Second Law is at least 100 times more powerful than the First.

For some reason, though, we haven't become as famous as we expected. It seems that most people claim to have known this law already, though nobody can give reference to a publication in a respectable journal. Perhaps the problem is that people *expect* too much of a Law. Barrington Moore seems to want an *explanation* of social continuity, not just a law proclaiming its universality. Well, that's not as easy, as it turns out. If we're going to *explain* why most of the time nothing much happens, it will take a whole book, and then some. But, if that's what we have to do to give Weinbergs' Second Law of Twins its rightful place, we'll just have to try.

The Continuity Taboo

What is human good? I think that it must be something more than sheer existence. For if sheer existence is the basic human good, then we might as well be oysters.

I submit that the human good includes at least the following two minimum components. The first is freedom from physical suffering, the most common forms of which are hunger and malnutrition, others being bodily harm due to natural disaster and illness. The second is freedom from boredom and psychological insecurity, especially fear of fellow human beings and crippling anxiety about one's place among them. (Hsu 1972:448)

It's easy enough to understand why "social continuity" sometimes requires bribery or judicious shooting, but why the reading of sociology? Perhaps we can illuminate this relationship by examining a passage or two from that subject, or from closely related subjects in the social sciences. We've chosen Hsu because we can honestly say we admire his work, for it represents much of what is best about sociology and anthropology.

In the passages above, Hsu wears his heart on his sleeve. Like most social scientists, Hsu is motivated by a deep concern for the quality of human existence. This concern leads him to insert *values* into his writing, something that is more rarely found in the sciences farther from the human condition. We could hardly imagine a geologist writing:

What is continental good? I think that it must be something more than sheer existence. For if sheer existence is the basic continental good, then continents might as well be islands.

For the most part, social scientists are in the employ of the established government, or of the "establishment." Given the nature of the selection processes through which one arrives at a secure position in government employ, only the rare social scientist will be deeply committed to social *discontinuity*. Like Hsu, they may wish for the common forms of suffering to be changed, but they will express their wishes in moderate ways—ways that inculcate the value of social continuity.

The history of American sociology is one demonstration of this compelling bond between employer and employee. Coser (1956) shows how the sociological study of conflict changed as sociologists ceased to be reformers and became tools of public and private bureaucracies. The reform-oriented sociologists of the early decades of the twentieth century considered conflict an inherent, even positive, feature of society. When conflict had negative consequences, a reworking of the social structure itself was necessary.

Contemporary sociologists, for the most part, regard conflict as necessarily disruptive, while society is static. Consequently, their focus is on conflict reduction and, along with it, the psychological adjustment of individuals to the social structure. Because conflict carries the potential for change, it is feared and avoided:

The decision-makers are engaged in maintaining and, if possible, strengthening the organization structures through and in which they exercise power and influence. Whatever conflicts occur within these structures will appear to be dysfunctional. Firmly wedded to the existing order by interest and sentiment, the decision-maker tends to view departures from this order as the result of psychological malfunctioning, and to explain conflict behavior as the result of such psychological factors. (Coser 1956:27–28)

If social scientists do begin to express the idea that social discontinuity might be desirable or, worse yet, begin to act on that idea, some of Barrington Moore's other maintenance mechanisms may be called into play. In the 1950s, when the committe for Un-American Activities was enjoying its power without glory, even the name "social science" was quickly replaced, in many quarters, with the term, "behavioral science"—hoping it would be less easily confused with "socialism." What better demonstration could be imagined that social scientists face social discontinuity with great courage—as long as it's necessary to preserve the greater continuity? And if the prerevolutionary Russian term for "social science" had been "capital science," no doubt it would have been judiciously changed to "social science." Indeed, many terms *were* changed after the Revolution, as they have been after *every* revolution.

One of the paradoxical aspects of human behavior, as opposed, say, to "continental behavior," is this sensitivity to words. Most social scientists are occupied full or part time in the education of youth. In particular, the smartest of the social scientists are engaged to educate the smartest of the youth, because the young of any society represent the greatest potential force for discontinuous change. Indeed, if it were possible to perpetuate the society without the discontinuous risk of creating a new generation every 20 years, someone certainly would have figured it out by now. The young, for one thing, don't know the meaning of important words, like "socialism" or "capitalism" or "freedom." They must be taught anew, each generation, or the meaning of those words might change, and, with them, the society.

But with the cleverest of youth, one must take care. If you should dare to *suggest* that they are sitting in a sociology class, reading sociology, in order to remove the threat of their creating a social discontinuity, *watch out!* No, it must be done subtly, without direct reference to continuity at all. And, if it can be expressed in terms of such stirring sentiments as "human good" and "freedom," the youth of the nation will usually gobble it up. By the time they are old enough to realize what has been done to them, they will be firmly rooted on the side of social continuity, ready to indoctrinate the *next* dangerous generation.

It's very difficult to talk explicitly about these matters, given the taboo we've just described. Somehow, we've learned, it's not *nice,* not academically *proper,* to question very deeply *why* things remain the same for so long, in the face of such discontent on the part of large segments of the society. When we speak clearly of them, it sounds as if we're being *critical.* Isn't this so? Didn't the preceding paragraphs sound disturbingly like some radical tract? Read them again and see if they contain anything other than factual material. They *sound* inflammatory only because of the unspoken taboo against speaking of such things in a factual manner.

It's relatively easy to speak in this way about the behavior of continents, or of machines—even of bacteria. When we get to the "higher" (i.e. closer to people) animals, the writing begins to sound political and may be attacked by the powers that be or the powers that aren't. Aesop and other political writers kept their heads by telling animal stories, but that strategy didn't always work. Just as with sex education, telling about the birds and the bees might cost you your job.

Social scientists, to summarize our argument up until now, are concerned with the human condition. They are also concerned with keeping their jobs and with ushering another generation safely through that dangerous age. They would *like* to speak about continuity, but cannot do so directly. Therefore, we contend, they have become extremely talented at doing so in more subtle ways.

Hsu's concerns—freedom from boredom and psychological insecurity—are typical concerns of behavioral scientists and of their bright students destined for responsible positions in the establishment. It's not clear that they are the principal problems of the great majority of ordinary people—the kind that don't have tenured positions in the established order to guarantee their bodily survival. Yet even from his ivory tower, Hsu has come pretty close to the mark. The minimum components he cites are all clearly and directly related to the problem of survival—and how we, as human beings, have been designed to solve it.

First, physical suffering—in the cases cited—is precisely an indicator of potential damage to survival chances. The suffering itself is a signal from our bodies to our brains that we are in danger of not surviving, a signal to get busy and do something about it, putting aside all lesser pursuits.

Second, boredom comes only because we are designed as anticipatory regulators, and because we've now done such a good job of regulating that we don't have to spend all our time at it. Nobody is bored on a whitewater canoe trip, or when lost in the rain forest.

Third, the psychological insecurity in us is quite well motivated by survival needs. On one hand, we require other human beings acting in society to maintain our unblemished survival record. On the other hand, we have carried the social strategy so far that the most likely source of nonsurvival today is other human beings. The contradiction creates the anxiety—another anticipatory signal that our regulation may be failing.

In other words, even if we want to go beyond "sheer existence," we find that we must deal with the way we are built with "existence" in mind. And not just us. Almost all of the "systems" we see around us at any moment are there because they solved the problem of existence for a time, at least. Or conversely, we rarely get to observe a system that hasn't managed to survive for a while.

Of course, questions beyond "sheer existence" are more interesting, at least to those of us who have rarely, if ever, had to wrestle with questions of survival. As we grow older, though, and the question of personal survival becomes problematic, our interest in the mechanics of continued existence grows. And as a society, when cracks in the seams begin to show—as they do now—we are more likely to listen to the preachments or give grants to the research of someone who claims to know something about patching.

But even if we're young, rich, and in love with the most transcendental ideas, we'll be binding our hands, covering our eyes, and clouding our brains if we aren't equipped with an understanding of how it is that some systems survive and others don't. That's why, though we, too, hope to push on to the more spectacular topic of how systems change and surpass themselves, we pause first to build a base of understanding of more mundane questions.

But does it require an entire book? Aren't most of the principles of stability obvious to anyone who cares to look? Perhaps, but there may be more to it than mere "looking." As Marc Bloch explains it:

For even those texts or archaeological documents which seem the clearest and the most accommodating will speak only when they are properly questioned. Before Boucher de Perthes, as in our own day, there were plenty of flint artifacts in the alluvium of the Somme. However, there was no one to ask questions, and there was therefore no prehistory. As an old medievalist, I know nothing which is better reading than a cartulary. That is because I know just about what to ask it. A collection of Roman inscriptions, on the other hand, would tell me little. I know more or less how to read them, but not how to cross-question them. In other words, every historical research supposes that the inquiry has a direction at the very first step. In the beginning, there must be the guiding spirit. Mere passive observation, even supposing such a thing were possible, has never contributed anything productive to any science. (1953:64–65)

The study of systems that survive is "history" in its most general sense. Therefore, the study of any such system—which means the study of almost any system we deem worthy of study—needs to be directed by a "guiding spirit" grounded in a knowledge of which questions to ask. This brings us back to the education of the young, for what behavior more succinctly characterizes the human pup than the asking of questions? What better way to produce social continuity than by teaching people which questions *not* to ask? We'll try to put them all in our study.

The General Systems Approach to Continuity

I agree entirely that, in this case, a discussion as to what is meant is important and highly necessary as a preliminary to a consideration of the substantial question, but if nothing can be said on the substantial question, it seems a waste of time to discuss what it means. These philosophers remind me of the shopkeeper of whom I once asked the shortest way to Winchester. He called to a man in the back premises:

"Gentleman wants to know the shortest way to Winchester."
"Winchester?" an unseen voice replied.
"Aye."
"Way to Winchester?"
"Aye."
"Shortest way?"
"Aye."
"Dunno."

He wanted to get the nature of the question clear, but took no interest in answering it. This is exactly what modern philosophy does for the earnest seeker after truth. Is it surprising that young people turn to other studies? (Russell 1956:169–170)

If we are to retain the interest of young people in the subject of continuity or discontinuity, we shall, according to Russell, have to do more than teach them what questions not to ask, or to ask. We shall have to attempt, at least, to lay down some *principles* of stability that pretend to answer these questions.

It will certainly be a good idea to start by getting the nature of the questions clearly in mind, but we mustn't go too long without any intention of answering them. Therefore, even though we're not always secure in the truth, generality, or applicability of our answers, we'll try to give *something*. And, as a test of that something, we'll try to phrase the principles in ways that make them potentially *applicable*, which is why we speak of *design* of stable systems.

But, really, aren't most principles of stability obvious, though perhaps neglected? Yes, they're obvious, at least in the sense that Newton's Laws of Motion or Darwin's Principle of Natural Selection are obvious. Once we bother to look closely, and actively, we can learn much in a short time. Afterwards, it will all seem obvious, and we will say, "Why, I knew that all along!"

But when you examine the writing and thinking of so many of our contemporaries, you find that few of them, apparently, have bothered to look closely and actively. Our entire book is filled with such examples. For now one almost universal example from the American press will suffice. In Western journalistic terms, Chinese culture is frequently portrayed in terms of an "anthill" analogy—a society surviving by sheer force of numbers. American society, on the other hand, is seen as rich in structure, with great value assigned to each individual part.

These two views—randomness and structure—represent the two polar, but complementary, strategies of regulation, so this common portrayal is of great interest to us. Yet if we examine the two societies closely, as Hsu (1972) did, we find that the truth of the matter is precisely the *opposite* of the commonly accepted (Western) view. Chinese society has far more structure, structure that ranges up and down the hierarchy and across the landscape. American society depends for its survival on a large number of individual actors all working as independently as possible, very much, indeed, like ants.

It matters, politically, which view is "correct," but for present purposes it matters little. Such a complete difference of understanding of the same facts means that we must be on to something. The subject must be worth studying.

Our study will, in fact, be organized along the lines of this great dichotomy between structure and randomness as strategies for survival. In social systems, biological systems, and artificial systems of all kinds, we see the contrast of these two fundamental principles of design for stability. On one hand, we have a collection of clearly differentiated parts, each with its own function and, on the other, a collection of indistinguishable, homogeneous parts, each performing one small fraction of the total function of the whole.

Durkheim, in *The Division of Labor in Society* (1933) called the one Organic Solidarity; the other, Mechanical Solidarity. Tönnies, in *Community and Association* (1955) burdened generations of English-speaking graduate students with Gesellschaft (Association) and Gemeinschaft (Community). Others have seen the same dichotomy in functional terms—Dependent versus Independent, Open versus Closed, or Fast versus Slow. It's our belief that all these views are parts of a unitary view of the process of surviving in a changeable environment. Consequently, we believe that the study of survival itself will prepare the reader for the study of many seemingly unrelated topics—in the sciences, in engineering, and in workaday life.

If we are correct, if we succeed in demonstrating these widely applicable thinking processes, then we will have fulfilled another small part of the program of the general systems movement. We will have made some of the most generally applicable insights available to the most general audience possible.

Please note well that we are *not* claiming some general theory of all knowledge, as many so-called "general systems" writers have done. All we claim is that there is a body of knowledge, much less widely known than it ought to be, that transcends many of the disciplinary boundaries in which parts of it have been confined. The previous volume (Weinberg 1975) explored the question of what we could know, generally, about systems by virtue of their being observed by us. To some readers, it ought to have been called a psychology book. To others, it was philosophy. To us, it took pieces from each field, pieces from other fields, and pieces too scruffy to be claimed by any field at all. Our guiding principle was, and is, "generally applicable insights available to the most general audience possible."

In this volume, we'll explore the question of what we know, generally, about systems by virtue of their having been around a while. Some people will call this physiology; others, sociology; still others, geology.

The more such claims, the more we will have succeeded. The previous volume has been adopted as a text in courses in geography, computer science, history, medicine, urban studies, metallurgy, theology, architecture, anthropology, art history, and business administration. We hope this one will succeed as well.

Whether or not this information is used to design stable systems or merely to understand existing systems, we believe the active exploration implied by "design" is appropriate. We are not speaking about a closed field, or even a field at all, in the usual sense. Our own interest in the general systems approach to stability problems comes both from its prescriptive (design) and its descriptive (understanding) power. We've applied, tested, and refined these insights in the design and development of computer systems, systems of education, and business and professional organizations. With the help of these insights, we've been better able to understand other cultures than those of our birth, other disciplines than those of our training, and other people than those we were taught to get along with. We've been able to communicate these insights to others in a classroom setting as well as in a consulting situation. Now we are using a different medium, one that allows of little of the feedback that encourages reliable communication. With the reader's help, however, we think it can be done.

QUESTIONS FOR FURTHER RESEARCH

1. *The Nobel Prize*

Relatively few scientists, and even fewer philosophers, receive the Nobel Prize, thus making the award a rare event, and potentially disruptive. List the stabilizing and disruptive effects that such prizes may promote. On the whole, are they a stabilizing or disruptive force in science and philosophy? Is that good or bad?

2. *Glory*

Putting someone's name on a "law" is another sort of prize. Compare this type of prize with the Nobel type, from the point of view of stabilization and disruption, and also from the point of view of good and bad.

3. *Philosophy and Revolution*

Lenin, in his essay, *What is to Be Done?*, said ". . . the role of (revolutionary) vanguard can be fulfilled only by a party that is guided by an advanced theory." Can a theory of stability be the basis for a party of revolution? Can a theory of stability be the basis for a theory of revolution? If you want to make a revolution, should you study stability or change? If you want to prevent a revolution, should you study change or stability? Can you study stability without, in effect, studying change?

4. *The Ethical Problem of Stability*

This notion (adjustment) is often left empty of any specific content; but often, too, its content is in effect a propaganda for conformity to those norms and traits ideally associated with the small-town middle class. Yet these social and moral elements are masked by the biological metaphor implied by the term "adaptation"; in fact the term is accompanied by an entourage of such socially bare terms as "existence" and "survival." The concept of "adjustment," by biological metaphor, is made formal and universal. But the actual use of the term often makes evident the acceptance of the ends and the means of the smaller community milieux. Many writers suggest techniques believed to be less disruptive than otherwise in order to attain goals as given; they do not usually consider whether or not certain groups or individuals, caught in under-privileged situations, can possibly achieve these goals without modification of the institutional framework as a whole. (Mills 1967:90)

How could a society be organized so that most social thinkers would advocate more disruptive, rather than less disruptive, techniques for attaining given goals?

5. *Biology and Values*

What's wrong with being an oyster, if you're an oyster?

6. *History*

What's a *cartulary*? Does it matter? Why? If it did matter, how would you find out what it is, and how to read it?

7. *Anthropology*

Jeremy Boissevain (1974) criticizes structural-functionalism as a theory of social order on the ground that this equilibrium model simply does not explain the realities of society. Yet, in spite of its explanatory inadequacy, the model has pervaded anthropological thinking for decades. "Entrenched power hierarchies in the scientific community thus inhibit the development of scientific theory," Boissevain writes (1974: 23). Study his explanations of "the persistence of a myth" (1974:19–23). Read *Reinventing Anthropology* (Hymes 1969) for more recent discussions on the relationship between scientific theories and the power hierarchies that sustain them and are supported by them. As you become a more proficient general systems thinker, develop an alternative model of social order and social change, keeping in mind both the scientific and political implications of your model.

2

Aggregates

The peasants who farm their own small holdings form the majority of the French population. Throughout the country, they live in almost identical conditions, but enter very little into relationships one with another. Their mode of production isolates them, instead of bringing them into mutual contact. The isolation is intensified by the inadequacy of the means of communication in France, and by the poverty of the peasants. Their farms are so small that there is practically no scope for a division of labor, no opportunity for scientific agriculture. Among the peasantry, therefore, there can be no multiplicity of development, no differentiation of talents, no wealth of social relationships. Each family is almost self-sufficient, producing on its own plot of land the greater part of its requirements, and thus providing itself with the necessaries of life through an interchange with nature rather than by means of intercourse with society. Here is a small plot of land, with a peasant farmer and his family; there is another plot of land, another peasant with his wife and children. A score or two of these atoms make up a village, and a few score of villages make up a department. In this way, the great mass of the French nation is formed by the simple addition of like entities, much as a sack of potatoes consists of a lot of potatoes huddled in a sack. (Marx 1957)

The First Law of Aggregates

> I wandered lonely as a cloud
> That floats on high o'er vales and hills,
> When all at once I saw a crowd,
> A host of golden daffodils . . .
>
> William Wordsworth
> from "Daffodils"

Of the two grand strategies for stability, *redundancy* is by far the more common—so common we balk at the term "strategy." Sometimes we need a poet to bring a commonplace to consciousness.

13

The feature that defines a redundant system is the *seeming excess of parts*. The system does not need so many parts to be recognized as a system. Wordsworth saw ten thousand daffodils at a glance, but would have recognized the crowd, the host, with far fewer than a thousand.

Because of this seeming excess, the system can survive the disappearance of one or more of the parts. Without redundancy, we could not pick a bouquet without obliterating the crowd, expunging the host.

How many daffodils may we pick without destroying them? The poet will not notice the loss of a bouquet but will be shattered if only a few stragglers are left. The flowers themselves, however, may successfully regenerate even from a single plant. Thus, the required redundancy depends on the purposes of the observer.

Consider the system that we call "a set of checkers." According to the rules, it must contain at least 12 red and 12 black pieces. If I'm afraid of losing some checkers and thus not being able to play, I can protect myself by obtaining a few "extra" pieces.

Suppose that you have a normal set of checkers at your home and a redundant set with 15 pieces of each color at your summer cottage. If you sometimes lose pieces, you can expect more summers than winters of checker playing. If each set loses one checker each year, you can expect to play only one year of winter checkers before the set is broken. The summer set will last at least four years, and more than that if the pieces lost are of different colors.

Notice how the survival time depends critically on the *rules of the game*. If I play backgammon—which requires 15 pieces of each color—my summer backgammon set will last but a single season. Yet the very same set of pieces—if viewed as a checker set—would last much longer. The dependency on the observer couldn't be more clear.

With games, there is usually a precisely specified critical minimum set of pieces, but most of our universal experience with redundant systems employs a much less precise limit. Indeed, when the number of "pieces" is sufficiently large, we may no longer perceive them as "pieces" at all, especially if they are small by human scale, as are grains of sand, or molecules. Nevertheless, the pieces are there, and if enough of them vanish, the system will vanish as well. Over centuries, a rock erodes and is gone. Some of its parts become grains of sand on the beach, but even the beach may disappear in a hurricane. Indeed, even the tiny grain of sand is an aggregate, capable of disappearing into its constituent molecules under a variety of conditions.

Any variable that owes its identity to a critical number of more or less uniform pieces, or parts, may be called an *aggregate variable*, even when the critical limit is not precisely defined. Just because we don't know which eroded grain makes the rock into a nonrock, we aren't prevented from dealing with the rock as an aggregate of sand grains. We must be cautious, however, when dealing with matters of exact survival times of such fuzzily defined aggregate variables.

Any system which we can usefully characterize by a single aggregate variable may be called, simply, an *aggregate*. When we think of aggregates, biological examples spring to mind: Hosts of daffodils. Swarms of bees. Flocks of sheep. Schools of herring. Blooms of plankton. Cultures of bacteria.

Inanimate aggregates also abound wherever we cast our eyes. A book is an aggregate—it survives the burning of its individual copies. An idea is an aggregate—it survives as its adherents are, one-by-one, burned, crucified, thrown to the lions, or destroyed by a Congressional Committee. A lake is an aggregate—it remains a lake even as millions of people flush billions of gallons of its water through their toilets.

A lake, or a sewage pond, is not a "pure" aggregate. Although one of its aggregate variables dominates our perception, the water supports many other systems. If we care to examine a book closely enough, we may find that it, too, consists of several different editions, or printings, aside from copies that have been struck incorrectly from the press, or have been modified by their readers. It's doubtful that any "idea" is the same in any two of its adherents' minds, but here, as with books and lakes, we can usefully *pretend* that they're all the same. Or so close that it doesn't matter, for whatever our purposes are.

Even with this pretense, pure aggregate systems are rare. Be it starling, starfish, star apple, or starflower, any multicellular organism has dozens of aggregate variables. In people, aggregate variables may be formed from the number of blood cells of each of several types, muscle cells, nerve cells, kidney cells, brain cells, liver cells, or even bacterial cells in the gut or on the skin. On a deeper level, the cells themselves may be characterized by hundreds of different aggregates of protein or other organic molecules.

As deep as we go, we find aggregate variables beneath other aggregates. A forest has aggregates of elms, oaks, and pines; each tree has aggregates of leaves and limbs, fruits and flowers; each fruit has an aggregate of seeds; each seed has an aggregate of cells; and each cell—like each cell in you or me—has aggregates upon aggregates of molecules.

Or again, a corporation has an aggregate of offices; each office has an aggregate of employees, of products, of records, of machines.

Whatever the aggregates, whatever the levels of aggregation, the essential significance for survival is the same. No aggregate is outlived by its worst member. This nearly tautological observation is the *First Law of Aggregates:*

> **When it comes to survival,**
> **aggregates outlive their worst members.**

The *chain* is *no stronger* than its weakest link. The *aggregate* is *no weaker*.

The First Law, though powerful in scope, is of little help in predicting how well the aggregate strategy works. Most aggregates far surpass the First Law's guarantee. A book survives as long as its *most ancient* copy. An idea is buried only with its *last* believer.

A lake lies somewhere between this unitary extreme and the limitation of the First Law. The lake's lifetime, assuming no replenishment, lies somewhere *between* the evaporation of the first and the last drop.

Because as observers we crave invariance, we frequently choose aggregate variables to identify our systems. The aggregate's lifetime lends our world the simplifying stability we need to live with the Square Law of Computation (Weinberg 1975).

We pay the price of simplification whenever it's important to discriminate individuals in the aggregate. The General watches without a care while his battalion goes into hopeless combat—until he recognizes his only son. Though *every* soldier is *somebody's son or daughter,* no General could be a General if he refused to think of his battalions as aggregates.

In the same way, no human could be human if *every* field of daffodils were ten thousand three hundred and seventy-two precious individual flowers. If it were, poetry—even thought itself—would be impossible.

Births and Deaths—The Fundamental Aggregate Equation

In 1966, the population of Italy was more than 53,000,000 inhabitants.

Tourist Guidebook

The tourist who meets the Venetian gondolier, the Roman driver, the Sicilian urchin, the Abruzzian novice, the Novaran stonecutter, or the Calabrian grandparent, will consider the guidebook description totally inadequate. Yet whatever touristic delights it omits, the guidebook contains at least one bon-bon for the mathematically inclined systems thinker.

All of the variability of the Italian population is summarized by characterizing the state of the "Italy" system by a *single number*—the "population."

Not that systems thinkers are insensitive to the richness of Italian life. Nor do they study aggregate systems *only* because they lend themselves to simple mathematical treatment (though that helps!). The truth is much simpler.

Aggregates are everywhere. Aggregates form the basis for the lowest levels of stability on which our entire world rests. They are, in a word, *general*. Thus laws governing aggregate behavior are *bound* to be relatively simple, general laws. For these reasons, most of the early systems researchers—whether or not they considered themselves systems researchers—were involved in the study of aggregate problems (Lotka 1924).

But even before the earliest systems researchers, mathematicians understood much of the behavior of aggregates. The *state* of a pure aggregate is described by a single variable standing for the *number of individuals in the aggregate*. "N" is the traditional name of this single variable, representing in many ways the simplest system we could possibly study.

An aggregate, being a system, is governed by the same functional relationship as any other system:

$$S_{t+1} = F(S_t, I_t)$$

(The state at time $t + 1$ depends only on the state at time t and on the input at time t, I_t.)

Because there is only one variable, N, in the system's state, the general relation becomes simpler and more specific:

$$N_{t+1} = F(N_t, I_t)$$

In plain English, *the number of individuals at one time depends on the number at the previous time,* and on any interactions with the outside.

If we want to study the behavior of N over time, we must examine two factors:

1. The rules governing *survival* of *existing* members.
2. The rules governing *creation* of *new* members.

As always, both factors depend on the observer's viewpoint.

In the customary terminology, a member who does not survive is called a "death." A newly created individual is called a "birth." There is no *necessary* connection between these terms and what we ordinarily con-

sider to be births and deaths. Thus a defecting Hungarian pilot para-
chuting into the Gulf of Trieste makes the population of Italy 53,000,001,
and a *bambino* born in Cravegna makes it 53,000,002. To a systems
thinker contemplating the fluctuations of the Italian population, both are
considered "births."

Similarly, when the Pope dies, the population is reduced to 53,000,001,
and it is further reduced to 53,000,000 when Maria Figliolia moves to the
Bronx to marry Antonio Ghilberti. Regardless of what the newspapers
think about the significance of these two events, both are simply "deaths"
to the systems thinker.

A birth, then, is the addition of an individual to the population, for any
reason. A death, similarly, is any sort of removal. Consequently, if we
know the number of births, B_t, and the number of deaths, D_t, that have
taken place between time t and time $t + 1$, we can compute precisely the
number of individuals present at time $t + 1$ from the number at time t.
All we do is make the general state equation specific by writing

$$N_{t+1} = N_t + B_t - D_t$$

This very specific way of describing the behavior of an aggregate system is called *The Fundamental Aggregate Equation*. It tells us that if we
can somehow describe B_t and D_t over some period of time, we can describe the behavior of N during that same period—provided we know the
value of N at the beginning.

Sometimes these descriptions are so easy as to be trivial. At other times,
they elude our most fervent efforts. For a given system, we don't know
in advance if we can succeed in pinning down the specifics we need, but
at least the Fundamental Aggregate Equation points us in the right direction.

In all work with aggregate systems, the first step is to break the problem into three main subproblems:

1.   Find the initial state, N_t.
2.   Find the sources of births, B_t.
3.   Find the sources of deaths, D_t.

For instance, if the population of Gambolia on 1 January 1941 is unknown
to us, we may be able to deduce it from other facts and figures more
readily accessible. First, we might try to find some earlier date on which
the Gambolian population was known. This will give us the initial N_t.
Looking in old yearbooks, we find that on 1 January 1940, the Gambolian
census yielded 43,821 inhabitants. Our first step is now complete.

Searching through hospital records, we find out that 94 Gambolians died in hospitals in 1940, and 71 were born there. Studying newspaper articles for 1940, we discover that two children were born in public places before their mothers reached the hospital, and that 21 people died in various accidents and homicides in the streets, at home, in the pub, or on the ski slopes. A visit to the immigration office shows that one Turkish family of four came to live in Gambolia during the year, along with the former Minister of the Treasury and her husband whose exile for crimes against the state had been lifted. At the visa office, we discover that 86 people actually left Gambolia in 1940, most of them to become streetcar conductors in the United States. Now, if our historical research has been complete, we can compute the population of Gambolia on 1 January 1941 from the facts that

$$N_{1940} = 43{,}821$$
$$B_{1940} = 71 + 2 + 4 + 2 = 79$$
$$D_{1940} = 94 + 21 + 86 = 201$$

so that

$$N_{1941} = 43{,}821 + 79 - 201 = 43{,}699$$

Although Gambolia is probably a fictitious country, historians do precisely this kind of demographic tabulation on real countries, cities, parishes, and neighborhoods. All the *hard work* is in the fact gathering, so it's good to be guided by a clear understanding of just what kinds of facts need to be gathered.

Unfortunately, many historians, sociologists, economists, biologists, and other scientists who study the behavior of various aggregates are not so systematic in their presentation of facts. They have spent so much time and effort gathering those facts that they understandably wish to enrich their writing with asides, footnotes, exceptions, qualifications, and curiosities. These make history and the like much more interesting than systems theory or mathematics, but they do sometimes confuse the reader who merely wishes to have a general idea of what's going on. Is the Gambolian population "withering away," as claimed by the opposition party, or is it increasing, as asserted by the royalists?

If a writer knits an overly intricate pattern of facts, we must unravel the pattern for ourselves, guided by the Fundamental Aggregate Equation. Sometimes this strategy actually reveals some strand the author forgot, or some dropped stitch. As we develop the subject of aggregate thinking, we'll see that everything starts from this one simple equation, even though the chains of reasoning can be quite remarkable.

Differential Equations and Chronological Graphs

The unifying principle of the analytic method has been the language of mathematics. Phenomena of widely different content . . . can be described by laws which have similar mathematical forms, namely partial differential equations of second order. In this way a common theoretical framework is provided for apparently unrelated phenomena, and consequently the principle of "economizing thought" is served.

As we have said, attempts to extend the analytic method to the study of living processes were only partially successful. . . . The reductionists assumed implicitly that if we knew enough about how living beings were put together, we could write down the equations that govern their behavior; and if we were clever enough in mathematics, we could solve the equations and so determine the "trajectories" of behavior. (Rapoport 1968:xv)

Not all the obscure writing in the world is done by humanists. Mathematicians and their disciples in science and engineering have been known to omit steps of reasoning from their printed arguments. If nonmathematicians are to follow this "economizing of thought," they are going to have to learn not to be intimidated by such common descriptions as differential equations. It's actually not very hard, if you know some tricks.

The simplicity of the Fundamental Equation makes it a valuable heuristic which will often enable a nonmathematician to read a paper that would otherwise be dismissed as hopelessly mathematical. Therefore, as an aid to the nonmathematicians, we must also take the trouble to disclose one more disguise in which our Fundamental Equation will be found.

We can, by simple algebra, move N_t over to the left side of the equation, giving

$$N_{t+1} - N_t = B_t - D_t$$

The left-hand side is now the *difference* between the present state and the previous state, and the equation has been transformed from one which describes the *state* to one which describes the *way the state changes* over one unit of time.

For the mathematician, it is a short step from this form to a form reading

$$\frac{dN}{dt} = B - D$$

which often befuddles the uninitiated. But the symbol

$$\frac{dN}{dt}$$

is just the mathematician's shorthand for saying "the rate at which N changes" and has nothing to do, as it might appear, with division or the symbol d.

Indeed, Isaac Newton, who invented the calculus while on a school holiday, used a different notation that is sometimes less confusing to beginners. To indicate the rate at which a variable changes, a dot is placed over the letter, so that

$$\dot{N} = \text{the rate at which } N \text{ changes (i.e., } dN/dt\text{)}$$

An adaptation of Newton's dot notation is to use a prime, as in

$$N' = \text{the rate at which } N \text{ changes (i.e., } dN/dt \text{ or } \dot{N}\text{)}$$

Each of the three notations has certain conveniences, but their continued survival in the literature of different fields is a vivid demonstration of the conservatism of science and scientists. If you want to swim in mathematical waters without being bitten by the sharks, you should learn to *read* each of them without fear—which will probably be easier than getting the notations changed to a single standard.

For instance, the Fundamental Aggregate Equation may be written as either

$$\dot{N} = B - D \text{ or } N' = B - D$$

either of which can be read in English as

the *rate* at which N changes is equal to the rate of births (B) minus the rate of deaths (D).

Because it deals with changes, or *differentials*, this type of equation is called a *differential equation*. Although mathematicians can perform spectacular stunts with differential equations, we need merely the ability to *read* them. Knowing only that much, we shall often find differential equations useful.

The relationship between the "rate of change with time" and a chronological graph is an intimate one, especially for a pure aggregate system. The graph represents the behavior of the aggregate over time, which is what we were trying to find out in the first place when we wrote down the Fundamental Equation. Therefore, the graph actually represents a "solution" of the differential equation. Contrariwise the equation summarizes

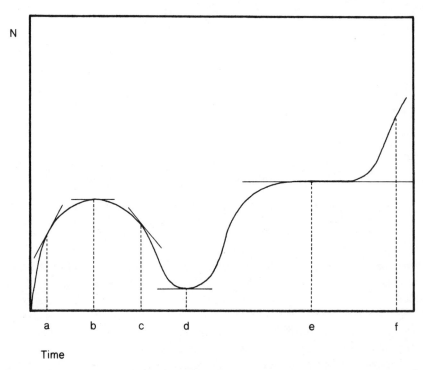

Figure 2.1. Slopes on a chronological graph: the slope is positive at a, zero at b, negative at c, zero again at d and at e. Where the slope is zero, the curve may be at a maximum (b), minimum (d), or point of inflection (e). Note that the "maximum" at b is smaller than the curve at f, making b a "local maximum."

the information in the graph. Learning to switch back and forth between the two representations enhances our systems thinking ability, for sometimes one form represents some fact much more clearly than the other.

For example, when N is plotted against t, N' represents the slope of the curve, or the instantaneous steepness. As we see in Figure 2.1, the slope at any point is the direction of a tangent line at that point. Five tangents are drawn on the chronological graph. The first, at a, is slanted upward—a *positive* slope—showing that the aggregate was increasing at time a. In equation form,

$$N'(a) > 0$$

which we read as "the rate of change of N at time a is greater than zero." This means, of course, that at time a the birth rate is greater than the death rate.

At time b, as the horizontal tangent shows, the death rate has caught up with the birth rate, so that they are just equal. That is,

$$N'(b) = 0$$

The population is neither increasing nor decreasing at that instant. It will soon be decreasing, as at time c, where the slope is downward, meaning an excess of deaths over births at that time. At d, the slope is again zero, but this time it is a minimum of population, not a maximum as at time b. In general, minima and maxima of N are characterized by the equation

$$N'(x) = 0$$

or simply

$$N' = 0$$

which tells us that on the chronological graph the slope is level. This might indicate a minimum or might indicate a maximum.

Quite often the minimum or maximum is precisely what we want to know about the aggregate. Such *extrema* can often be found without solving the differential equation just by setting $B = D$ and solving by simple algebra, or by plotting a few tangents to get a general idea of the curve. There are, however, some cautions that must be observed. First, we cannot tell just by $N' = 0$ whether we have a maximum or a minimum, but we can usually find out by sketching a bit of the chronological graph. We must be sure to sketch on both sides of the time in question, for we may also get zero slopes at points such as e, which is neither maximum nor minimum, but merely a "point of inflection" where the curve is, in this case, merely "resting" in a continued upward growth pattern.

We must also be wary of concluding too much from finding a maximum or minimum by this method, for it only locates what is called a "local" maximum or minimum. At b, for instance, the aggregate is clearly larger than before or after; but later, at f, it becomes much larger than at b. In other words, if $N' = 0$, we may be at the top of a hill, but not necessarily on top of a mountain. To return to our touristic example, if the guide tells us that the slope of the Italian population was zero in 1966, and, even if we know that it was a maximum at that time, we may not safely conclude that it will be smaller than 53,000,000 in 1976. But if the guide tells us that the slope is positive in 1966, and if it remains positive through 1976, then we do know that the population will become greater than 53,000,000.

We do not need to solve differential equations to draw such conclusions. We need only keep in mind that Italy is being treated as a simple aggregate consisting of many more or less similar members—"a lot of potatoes huddled in a sack," to borrow Marx's description. Having decided that redundancy characterizes this system, we can observe it simply by noting how its size changes over time.

QUESTIONS FOR FURTHER RESEARCH

1. *Gerontology*

The amount of loss that various aggregate systems can tolerate and still function varies from almost zero to almost 100 percent. In the organs of the human body, for instance, the amount seems to vary between perhaps 20 and 60 percent. In ancient armies, the amount was evidently less than 10 percent, since to "decimate" a force came to mean not only the loss of one-tenth, but also the effective dissolution as a fighting unit. Make a list of simple aggregate systems and try to estimate the functional cutoff level for each. Then see if you can discover general principles for different classes of systems, such as organs, human organizations, groups of animals, physical structures, or electronic equipment. (References: Shock 1957; Strehler 1959)

2. *Numismatics*

Coins make interesting aggregates for study because they are marked with their date of birth. (Occasionally, like people, they falsify their dates, which might have to be taken into account.) Assemble as random a collection of one type of coin as possible. For example, you might buy all of the nickels or dimes that are removed from a vending machine over several weeks. Study the distribution of ages of these coins and try to make inferences about the death processes which coins undergo. (Reference: Kosambi 1966)

3. *Oversurvival*

Like any regulatory strategy, the strategy of aggregation can be overdone. In the individual body, we call this condition obesity; in saving, miserliness; in business, excess profit; in species, overpopulation; in war, overkill. Discuss what strategies a system can adopt to prevent oversurvival, which might by itself result in death.

3

Birth-Free Aggregates

No practical biologist interested in sexual reproduction would be led to work out the detailed consequences experienced by organisms having three or more sexes; yet what else should he do if he wishes to understand why the sexes are, in fact, always two? The ordinary mathematical procedure in dealing with any actual problem is, after abstracting what are believed to be the essential elements of the problem, to consider it as one of a system of possibilities infinitely wider than the actual, the essential relations of which may be apprehended by generalized reasoning, and subsumed in general formulae, which may be applied at will to any particular case considered. (Fisher 1929:ix)

Making reproduction more complex is only one way the mathematician works. Another approach is to make reproduction simpler—even eliminate it altogether. What we can learn by this technique might not be more interesting than the study of three sex systems, but it may prove more useful.

Social Versus Innate Survival

> For God's sake, let us sit upon the ground,
> And tell sad stories of the death of kings:
> How some have been deposed; some slain in war;
> Some haunted by the ghosts they have deposed;
> Some poisoned by their wives; some sleeping killed;
> All murdered:—for within the hollow crown
> That rounds the mortal temples of a king
> Keeps Death his court; and there the antick sits,
> Scoffing his state, and grinning at his pomp;
> Allow him a breath, a little scene,
> To monarchize, be feared, and kill with looks;
> Infusing him with self and vain conceit—
> As if this flesh, which walls about our life,

25

Were brass impregnable; and humored thus,
Comes at the last, and with a little pin
Bores through his castle-wall, and—farewell king!

William Shakespeare
King Richard The Second

Much systems writing uses the term "population" as we have used the term "aggregate." In the modern world, "population" is a term heavily laden with political connotations, so we shall reserve "population" for one subtype of all possible aggregate types.

A population is an aggregate with the ability to determine *its own birth rate through self-reproduction of its members.*

Not all aggregates are populations. In the most frequently occurring aggregates, there are *no new members at all*—after the creation of the initial aggregate. Rather than plunging into the complexity of populations, we will do well to work our way up from simpler cases. In many ways, these "birth-free" aggregates are the simplest of all.

A birth-free aggregate cannot increase in size.

In our fundamental aggregate equation,

$$N' = B - D$$

if $B = 0$, then

$$N' = -D$$

and the rate of change can never be positive. The best the birth-free aggregate can do is remain static—and that only if all of its members survive indefinitely. If $D = 0$, then $N' = 0$.

Examples of birth-free aggregates are supplies of nonreplenishable ores, human brain cells or nerve cells (according to one theory), and machines that are not maintained. Other examples are innumerable collections of items that are no longer produced, such as Gutenberg Bibles, American silver dollars or gold double-eagles, upside-down airmail stamps, Spanish-American War veterans, or Cremona violins. Although all such birth-free aggregates must *inevitably* disappear, the duration and manner of disappearance vary greatly according to the underlying mechanism.

In the simplest such system, each member has a predetermined time of death. If we know all these times, we can plot the decline of the birth-free aggregate in advance. For instance, if we issue 3,000,000 credit cards, with 250,000 due to expire each month of the coming year, we know that after 12 months, no valid cards will remain. This is precisely the technique credit card companies use to limit their liability for lost or stolen cards.

In other birth-free aggregates, we may not *know* exactly when each *particular* member is destined to die. In an elimination tournament involving 128 teams, we know that 64 will "die" on the first round, though we don't know *which* 64. We also know that 32 will die on the next, 16 on the next, 8 on the next, 4 on the next, 2 on the next, and 1 in the finals. After that, there's no more "tournament," but only a "winner."

The entire life insurance industry is predicated on precisely an ability to predict human deaths in a given age group *statistically*. An insurance company need not know *which* of its policy holders will die in a particular year. But in a great natural disaster, when many *correlated* deaths occur, the simple life insurance assumptions fail—along with insurance companies that counted too heavily on the independence of its policy holders' lifetimes.

Disasters aside, human deaths are quite predictable—if you know a few significant facts about the individuals in the aggregate in question, such as age and sex. We can even imagine a simpler situation, in which *no* facts about the individuals make any difference to the chances of death. In such a system, every living individual has the same chance as every other of dying (or surviving) in any period of time. As we shall soon see, this special simple case yields a death pattern we call _exponential decay_.

For the behavior of a birth-free aggregate to differ from the behavior of an individual member, there must be some *variation* in the individual behaviors. A common situation is shown by the aggregate whose individual members are the lighted lamps along a street or on a Christmas tree. Although there are any number of ways in which the lamps might be wired to the central power source, two particularly common and simple ways serve to illustrate our point—series and parallel wiring, as shown in Figure 3.1. From the point of view of survival, the parallel arrangement is superior to the series arrangement. As soon as one of the series lamps burns out, the single circuit for all the lamps is broken. They all go out, leaving the street—or the Christmas tree—in darkness. In the parallel connection, on the other hand, one burned out light breaks only its own part of the circuit, leaving the other lights undisturbed. Figure 3.2 shows chronologically the performance of the two types of wiring, clearly il-

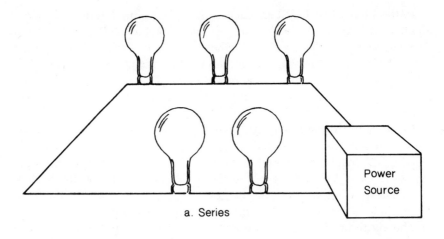

a. Series

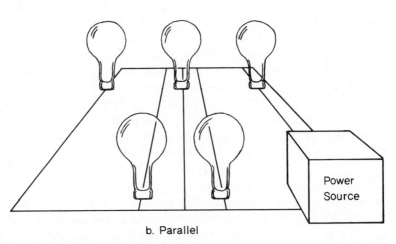

b. Parallel

Figure 3.1.　a. SERIES
　　　　　　　b. PARALLEL
In the series structure (a), current passing through one bulb must pass through all bulbs, so if one fails, all fail. In the parallel structure (b), each bulb takes a fraction of the total current, so one bulb failing need not affect the continued lighting of another.

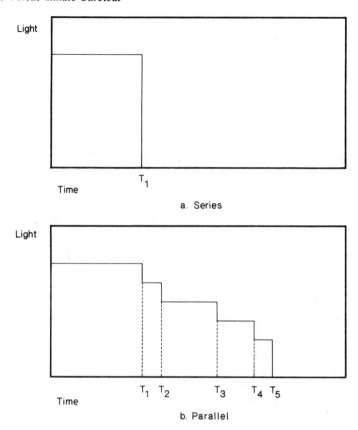

Figure 3.2.   a. SERIES
              b. PARALLEL
Chronological graph a can represent the failure of one bulb in a series structure causing all bulbs to stop receiving current. It could also represent the behavior of a paral-
lel structure in which all the bulbs have exactly the same lifetime ($T_1 = T_2 = T_3 = T_4 = T_5$), which changes graph b into graph a.

lustrating how the series connection provides light only for the lifetime,
T_1, of its shortest-lived bulb, while the parallel connection provides some
lighting as long as its longest-lived bulb, T_5, still shines.

Suppose, however, that all bulbs had exactly the same lifetime, that is,
$T_1 = T_2 = T_3 = T_4 = T_5$. In that case, the parallel connection would provide no more dependable light than the series connection and the chronological graph for parallel wiring would look exactly like Figure 3.2a,
for series wiring. We would be unable to distinguish from black box observation whether:

1. The bulbs were connected in parallel with identical survival times.
2. The bulbs were connected in series.

Even in this simplest of all examples, we see that the behavior of an aggregate is influenced by both

1. The way one member's survival *depends* on the survival of the others.
2. The way one member's survival varies *independently* of what the other members are doing.

To remember this important dichotomy, we can think of (1) as the *social* component of survival and (2) as the *innate* or individual component. You may find it helpful to call (1) *congenial* behavior and (2) *congenital* behavior, but they're awfully confusing in print.

We'll stick with "social" and "innate," but not without some fear of getting involved in heated political controversies over which parts of our behavior come from our society and which parts we're born with. According to Shakespeare, all kings are murdered, which is certainly a *social* cause of death, but they become kings by accident of birth, which could be considered *innate*, though cultural.

For now, just consider "social" and "innate" as definitional words, like "births" and "deaths." Inasmuch as we've now seen that innate and social behavior cannot always be distinguished by the black-box observer, we'd be wise to temper our involvement in such monarchizing.

Exponential Decay

> Why did you die when the lambs were cropping?
> You should have died at the apples' dropping,
> When the grasshopper comes to trouble,
> And the wheat-fields are sodden stubble,
> And all winds go sighing
> For sweet things dying.
>
> Christina Rossetti
> *A Dirge*

To keep matters simple, let's first consider some cases in which there is no "social" behavior—where the death of one individual depends in no way on the life or death of another. The death of lovers often appears that way to us. Had they any consideration for the living, they would have died at a more appropriate time.

Actually, there's some doubt that human deaths are so random. It's been observed that in nursing homes, people rarely die in the month before their birthday, as if they willed themselves to live for just one more year. Let's avoid the sadness and uncertainty of actual death and consider cases that are merely mathematical deaths.

A village of 1000 adults has been chosen as the location for an opinion poll. Each week, a poll is taken, during which 100 adults are chosen *at random*. In the first poll, none of the 100 will have ever been polled before, and 900 will remain as people "not yet polled"—as suggested in Figure 3.3.

In the second week, we may happen to poll some people we have already polled the first week. Indeed, if we are choosing at random each week, approximately 10 of the 100 people selected in the second week will have been selected in the first poll. This time, then, only 90 additional people will be removed from the list of people "not yet polled," since 10 had already been removed.

Now suppose we are interested in the aggregate of people "not yet polled." We might want, for instance, to have some idea of how many people have *never* contributed an opinion to our surveys. For one thing, after several weeks we can check this collection of not-yet-polled people for any sign of possible nonrandomness, to soften criticism of our polling methods.

The not-yet-polled aggregate is birth-free, if we assume nobody is moving into town, or if we're using a fixed list of the 1000 official inhabitants. After the first poll, its N decreases from 1000 to 900. After the second, we can only predict the decrease statistically, but it will probably drop to about $N = 810$. By similar reasoning we see that it then decreases to 729, 656, 590, 531, 478, and so forth. The decline in this not-yet-polled aggregate is a classical example of *exponential decay*.

If the initial aggregate—called N_0 ("N at time 0") is amply proportioned, the exponential decay can be represented by the smooth curve illustrated in Figure 3.4. It may represent unpolled people. It could equally well represent unkilled flies remaining on a screened porch. For this model to apply to the screened porch, we need a constant swatting (polling) effort, the porch (town) must be sealed so no flies (people) enter or leave, and the eradication must be sufficiently fast so that flies neither breed nor die of old age during the observation period.

Any system can be modeled by exponential decay, if it is a collection of individuals who cannot die of old age, but who are exterminated by random "accidental" causes. Some electronic components in computers

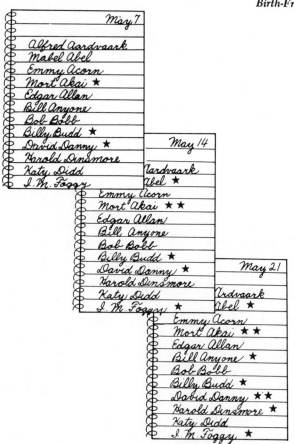

Figure 3.3. We can keep track of who has been polled by pasting a star on the village census next to the name of each person polled that week. There will be 100 stars pasted the first week, but in the second week, about 10 of the polled names will already have stars from the previous poll. The unstarred names are the people not yet polled at any time.

seem to behave in this fashion, once some nonrandom initial failures have been weeded out. Another such collection is the atoms in a lump of radioactive material—unless the lump is of critical mass, in which case they are very social and all decay in a very short time. We call *that* system an atom bomb—which is a case of exponential *increase*, rather than exponential decay.

The curve of exponential decay has a *continuously decreasing* slope as the population gets smaller. In fact, the true mathematical exponential decay curve gets so flat in its tail that it never actually reaches zero. This would mean that exponentially decaying aggregates would never truly

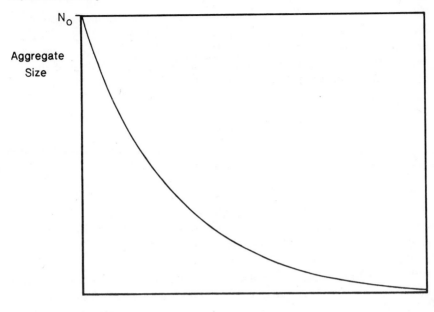

Figure 3.4. Exponential decay of an aggregate initially consisting of N_0 individuals, where N_0 is large enough for the curve to look smooth. The curve can represent many aggregate systems, including "the number of people not yet polled."

disappear, were we to forget that the ideal curve only *approximates* a finite aggregate.

The exponential decay curve is derived from the assumption that a *constant fraction* of the population is removed at each interval of time. When the aggregate is still quite large, this assumption may be quite a close approximation to its behavior. But when there are no more than a few members left, the approximation gets impossibly poor.

In particular, when a *single* member is left, it either disappears during a given time period or it doesn't—so the fraction must be either one or zero. This is an example of an "end effect." If we want to know the exact behavior of a system at *either end* of its life span, we must examine individual cases. In this case, we may apply *The Law of Small Numbers:*

Theories built on large number approximations usually don't work when applied to the behavior of small aggregates.

Notice, however, that we said, "If we want to know the *exact* behavior . . ." There are many interesting questions about exponentially decaying systems that don't require such exact information. For instance, radioactively decaying elements, such as carbon-14, can be used to determine the age of archaeological artifacts. How is this done, when the exponential curve cannot give us a precise limit on the lifetime of the aggregate of carbon-14 atoms in a piece of wood? The answer is that we are not seeking the lifetime of the *carbon* but of the *artifact*. For the method to work, in fact, the carbon-14 aggregate must *not* yet be exhausted.

From a sample of the material in the artifact, the amount of carbon-14 relative to nonradioactive carbon-12 is determined. Assuming that the *original* ratio of the two carbons is known, we can determine how much of the original carbon-14 has died. We can then use the concept of "half-life"—the time it takes for the original aggregate, N_0, to fall to half its size, $N_0/2$—to determine how long the radioactive decay has been going on. If no appreciable amount of carbon-14 has been added to the artifact since it was made, we then know that the time of decay is the age of the artifact.

The half-life (which we will designate by H, rather than some more conventionally used symbols), is illustrated in Figure 3.5. The figure also shows that the aggregate is halved again, to $N_0/4$, after another H has passed. In fact, as the reader may verify, the aggregate size continues to be halved with the passage of each H. From any point on the curve, the point H in the future will be just half as high, and the point H in the past will be twice as high.

Suppose, then, that we have a carbonaceous artifact with carbon-14 measured at $1/8$ of the original amount. This is derived from knowledge of the ratio of the two carbons in the atmosphere. If all our simplifying assumptions are correct, or can be corrected for, then there have been three half-lives ($1/2 \times 1/2 \times 1/2 = 1/8$) since the sample stopped adding carbon-14. Knowing that the half-life, H, of carbon-14 is 5700 years, we conclude that the sample is $5700 \times 3 = 17,100$ years old. Of course, this is only an approximation, but it's probably good to 1000 years one way or the other. This might not sound very accurate, but it's far more accurate than we could predict the exact time it will take for the *last* carbon-14 atom to decay.

The half-life, then, gives us a useful tool for thinking about the way an aggregate decays. It also enables us to construct a neat exponential curve with pencil and ruler when the only information available is N_0 and H. We can also use the half-life to make quick but educated predictions about the future behavior of some aggregates. We can, for example, keep our supply of spare electronic parts at a comfortable level without wasting money on oversupply.

Actually, the half-life is merely one conventional way of characterizing

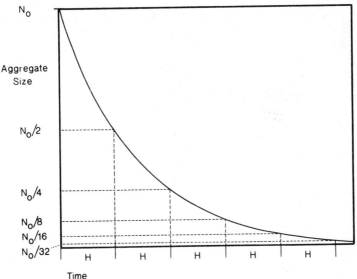

Figure 3.5. An exponential decay curve created using the concept of "half-life"—the time it takes for the aggregate to be cut in half.

exponential decay. We could as well have used the quarter-life, or three-quarter-life, or 38.2-percent-life, for that matter. The simple fact is that the exponentially decaying birth-free aggregate declines at a *constant rate* over *constant intervals of time*. If the time interval is longer, the rate per interval is larger; if smaller, the rate per interval is smaller. Our opinion pollsters removed 10 percent of the not-yet-polled each week, so the 90-percent-life of this aggregate is exactly one week. That is, in one week's time, its numbers are reduced to 90% of its previous size. This corresponds to a half-life of about 6½ weeks, not 5 weeks, as we might erroneously reason by thinking, "Well, 10 percent per week is 50 percent in 5 weeks."

Exponential decay often fools people who don't understand that the death rate is a fixed *proportion of the aggregate* at any moment, rather than a fixed *number of members*. That is,

$$D = pN$$

where p represents the constant rate, or proportion, of depletion. In the opinion poll, $p = .10$, for the time interval of one week.

Now let's compare this fixed *proportion* depletion with fixed *number* depletion, which would be

$$D = pN_o$$

Suppose the pollster lays in a supply of imported cigars when he comes to the village, knowing that the local tobacconist does not carry them. Because he *is* a heavy smoker, the pollster notices that he smokes exactly 10 percent of his initial stock of 1000 cigars in the first week. "Aha," he reasons, "I polled 100 people out of 1000 and smoked 100 cigars out of 1000. Therefore, I'll have just enough cigars to last until everybody has been polled!"

Well, what do you expect from someone stupid enough to smoke 100 cigars a week?

Alternative Ways of Representing Decay

> How well the skilful Gardner drew
> Of flow'rs and herbes this Dial new;
> Where from above the milder Sun
> Does through a fragrant Zodiack run;
> And, as it works, th' industrious Bee
> Computes its time as well as we.
> How could such sweet and wholsome Hours
> Be reckon'd but with herbs and flow'rs!
>
> Andrew Marvell
> *The Garden*

The half-life concept can be used to measure decay, or, turned around, to measure the passage of time. If "herbs and flow'rs" die exponentially, we, like the bee, could compute time directly from the garden. Seeing the decay in terms of half-lives gives us reasoning powers that can be applied to practical, as well as poetical, situations.

The half-life concept helps us to develop *intuition* about the meaning of exponential decay, and exponential decay itself helps us to develop intuition about other forms of death curves. Anyone working extensively with aggregates needs to develop such reliable intuitions, by whatever means possible. Although "herbs and flow'rs" might not always be available, our intuition functions best when it can be based on natural, or nearly natural, phenomena. If, for instance, we can harness our *visual sense* to aid our verbal reasoning, we will be drawing upon hundreds of generations of accumulated "wisdom of the eye." Therefore, let's see what we can accomplish with some simple graphical transformations.

We can transform the differential form,

$$N' = B - D$$

into a specific equation that incorporates these two properties of exponential decay:

1. Zero birth rate ($B = 0$).
2. Proportional depletion or death rate ($D = pN$).

giving

$$N' = 0 - pN = -pN$$

For any time, t, then, we can use this equation, $N' = -pN$, as we used the half-life concept—to give us the current value of N and to plot a curve which predicts all values of N.

This predictive use of our transformed aggregate equation can be checked by recalling our interpretation of N' as the *slope* of the curve. If the slope is $-pN$, then the larger N is, the steeper the negative slope will be. In other words, the larger the aggregate, the faster it is decreasing, which can be verified by moving a rule along the curve as a tangent in Figure 3.4 or 3.5. The ruler will become progressively more horizontal as time increases—that is, as it is moved to the right.

The need for a ruler is a clue that it's not easy for us to follow such a curve accurately with the unaided eye and brain. Our intuition might be happier with a *straight* line. After all, that's what the cigar-smoking pollster thought he had—and did have, for his cigars.

The more intuitive form can be achieved by a common plotting trick. Instead of plotting constant *numbers* remaining in N, we can plot in terms of constant *proportions* remaining in N. The same size interval that represents our pollster's 1000 to 500 ($N/2$) is used to plot the interval from 500 to 250 ($N/4$), as shown in Figure 3.6. Each fixed *interval* thus represents a fixed *proportion*, but a constantly decreasing *number*—$N/2$ ($N = 500$), $N/4$ ($N = 250$), $N/8$ ($N = 125$), and so on. This type of plotting is called *logarithmic plotting*, or log plotting.

By plotting log of N against time (T), we get constant declines for a constant move to the right—a straight line for the exponential decay curve. This type of plotting particularly facilitates *comparison* with other forms of aggregate decay, because we can so easily recognize deviations from a straight line. And comparison is the major reason we use log plotting, as we shall see.

Log plotting is but one of many different ways of displaying the behavior of systems in forms that appeal more directly to human intuition. For the inexperienced, however, the introduction of another form of plot seems to make understanding more *difficult*. Whenever you encounter an unfamiliar plotting scheme, keep in mind that *someone* designed it to

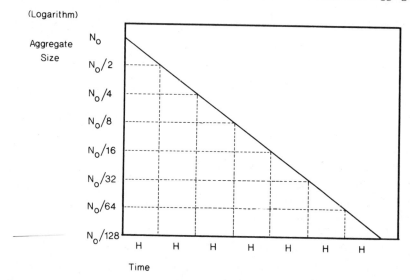

Figure 3.6. When plotted on a logarithmic scale, the exponential curve becomes a straight line, which is a convenient form for identification and comparison with other curves.

make *something* easier to see. Also keep in mind that a special plotting scheme is just like special *language*. It has a tendency to facilitate communication among insiders at the cost of erecting a barrier to communication with outsiders. Log plotting is no exception. If our purpose were merely to teach a few things about exponential decay, we would probably choose to live without log plotting in this book. But our purpose is much broader. We want to train systems thinkers, people who are unafraid to cross whatever barriers lie between the traditional disciplines. Therefore, we've introduced log plotting to let you cut your teeth, if you're unfamiliar with such plots, and prepare you for the harsh world of disciplinary reality. For those of you to whom this kind of visual presentation was mastered long ago, please be patient with your less fortunate colleagues.

Even the experienced ought to obey a few cautions when using any sort of graphic scheme. Logarithms, for instance, may tend to deemphasize fluctuations in data. By plotting logarithmically, we may be throwing away some important bend or wiggle without noticing.

Another misimpression that the log plot of Figure 3.6 may give is that the exponential curve has gone to zero, when in fact it *never* goes to zero. The bottom axis in the plot is $N/128$, or about .01 of the original aggregate. For some aggregates, .01 would be sufficiently close to zero to suit our purposes, but it depends on both the aggregate and the purposes. In some cases, it is precisely this tiny remainder that most concerns us.

For example, certain insect eradication strategies are quite similar to our previous model of eradicating flies on a screened porch. Advocates of this strategy—they may, for example, be selling insecticide—will produce curves looking like Figure 3.6. Perhaps they sincerely believe that their product leads to total eradication of the insect. The line certainly *seems* to go to zero.

Actually, of course, it goes only to some small fraction, and we may be unhappily surprised when next season the insects are back in full force— all descended from the "zero" survivors. Any representation of a system tends to make certain insights easier—at the expense of making others harder. Be sure the representation you're using isn't hiding the one thing you most want to see.

Another common way of looking at dying aggregates is not to plot N at all. Instead, we plot N', the *rate* at which N is changing at any particular time. This is easy to do in the case of the exponential because

$$N' = -pN$$

Except for the scale factor, $-p$, the curve for N' looks exactly like the curve for N. In other words, the *rate of dying* decreases exponentially along with the total number of survivors. Most of the aggregate die as infants; few die as patriarchs.

The plot of N, or log N, versus time may be thought of as showing the chances of any individual in the original aggregate *living* to a certain age. Since H is the half-life, half will die at or before time H, and half will live at least that long. Only one-fourth will live to age $2H$, and one-eighth to age $3H$.

The plot of N', on the other hand, (Figure 3.7) shows the chances of any individual in the original aggregate *dying at* a certain age. For the exponential, the chances of dying young are very great. For each older age, the chances of dying at that age are smaller and smaller—when considered at the very beginning.

The plot of N' versus time is called the *life table*. Actually, *death table* would be a more descriptive term, but life table is more in keeping with the terminology of the death insurance industry. Whoops, that's "life" insurance, as in "life" table.

Whatever it's called, the life table can be most useful in distinguishing among different types of aggregates. Because it's a plot of the *rate of change* of death, N', it emphasizes changes. By accentuating the bumps and wiggles, it allows us to see the differences we consider important between two similar aggregates.

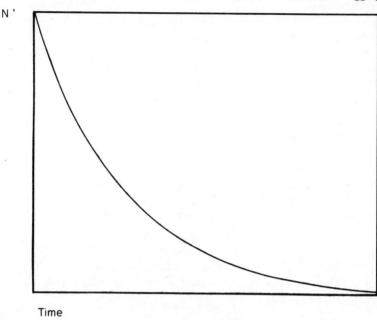

Time

Figure 3.7. The plot of N' for the exponential curve has the same shape as the exponential curve itself, showing that the probability of an individual dying young is very great.

Life tables may be given to characterize any aggregate, not just those that are birth-free. When there are births, the life table is usually given as the probability of one individual dying at a given interval after birth. This form is perfect for life insurance, and generally seems a very "natural" way of characterizing the death pattern of an aggregate.

Unimodal Life Tables, and Ogives

> The falling leaves
> fall and pile up; the rain
> beats on the rain.
>
> Gyodai
> *Autumn* (A Haiku)

Leaves do not fall exponentially, nor does the rain. First they fall, then they pile up, then a few stragglers fall after winter has already begun. A

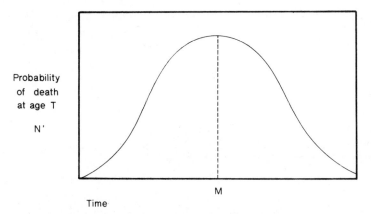

Figure 3.8. The life table for a set of "similar" parts, which last an average of M years before failing. A few fail very early, and a few fail very late, but most are just about average.

single drop on your face warns of the impending shower. Then the drops are several. The rain beats upon the rain, and then a few drops trail all the others. In our experience of the world, such distributions are far more familiar than the exponential, useful as it may be as a limiting case.

From the life table of such an aggregate, we readily see that *most* members are *pretty much* the same with respect to death, but not exactly the same. And, there are those few early drops, those last tenacious leaves, which always defy easy characterization. A plot of the life table for such an aggregate is shown in Figure 3.8. It could represent the leaves over a few weeks, the rain over a few hours, or us, over a few score years.

The life table of Figure 3.8 could represent a mechanical part designed to last for about 20,000 hours of operation. No matter how careful the design and manufacture, not all copies of the part will last exactly 20,000 hours. For a variety of reasons,

1. Some will die very young, but not very many.
2. As the operating time nears 20,000 hours, the probability of failing will increase, peaking at just about 20,000 hours.
3. There will be a few that last much longer than 20,000 hours for some reason nobody can exactly pin down.

Figure 3.8 could be the familiar "normal distribution," or it could just resemble a normal distribution with a single peak tapering off on both sides. Automobiles of a given model year obey a life table something like this, though clearly they can't have a normal distribution of lifetimes since the normal distribution is *symmetrical,* while cars may have an average life of 5 years and a longest life of 50 years or more. When new, they aren't likely to be scrapped unless they are involved in a wreck. Eventually, however, they begin to wear out, rust, and come loose in the joints. More and more are lost, peaking at some typical life for that model, until they become sufficiently rare for antique car buffs to start preserving them.

Another aggregate with this sort of death pattern might be the set of people who haven't yet heard the latest rumor. Another might be the trees in a forest that still have their leaves on a particular autumn day. With rumors, a few people are "always the first to know," while spouses are "always the last," at least for certain rumors. Most people, however, hear the news at about the same time, and this modal time would be easily spotted in a curve such as Figure 3.8, which equally well describes the life of a collection of identical parts, the set of all cars of the same model, the still-clad trees in the forest, the spread of a rumor or a joke, or a thousand other aggregate phenomena.

Because these life tables have a single peak, mathematicians sometimes refer to them as *unimodal* distributions. In Latin this just means they have one peak. The life table for the exponential decay really had no peak, or perhaps we could say it had a peak at time zero. Other aggregates have bimodal or more complex life tables, but for the moment we still have much to learn about unimodal patterns of dying.

A unimodal distribution cannot be completely described by a single number, like the half-life, H, for exponential decay. If the peak of the life table is sufficiently sharp, however, the modal value, M, can provide a rather accurate description. We can sketch a reasonable approximation just by knowing the location and value of M, since it is the place where the aggregate size is falling most rapidly. Both before and after M, the slope of the N curve will be flatter than at M itself.

One way to draw the N curve is to start with a sketch of the N' curve, since N' is precisely the slope of the N curve. We have to know N_0, if we want to put a scale on the N curve, but if we're just studying aggregate types, we can ignore the particular value of N_0, since big or small aggregates will display the same *pattern.* In Figure 3.9, we see the N curve constructed from the N' curve of Figure 3.8. It looks like a backward S, and is often called an "S curve." Another name for this or its reverse curve is "ogive," (pronounced "Oh, jive!") from an obscure French word used to describe an arch made of two such curves attached together.

The term ogive is also used to describe *any* N curve derived from a corresponding N' curve for a birth-free aggregate, perhaps under the mistaken impression that all such curves must be S-shaped. Later, we shall

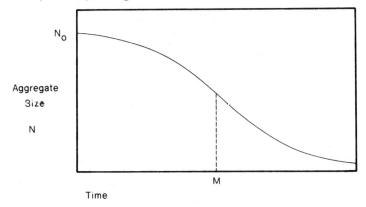

Figure 3.9. The number of non-failed parts still in operation after a given period of time forms an *ogive* when the life table for the individual parts is similar to that of Figure 3.8.

see examples that don't follow this "rule"—an unfortunate case of people coming to take the limited world of their mathematical models for total reality.

The N curve, whatever its ultimate shape, describes *change* in aggregate size, and is thus simply a transformation of the N' curve, which describes the *rate* of that change. Those with training in calculus will be able to see that the N curve is the "integral" of the N' curve. Those without benefit of such training will be able to think about the matter and soon see that the N curve represents the sum (turned upside down, to be sure) of the deaths so far, as predicted by the N' curve, or life table.

If the N' curve is expressed in discrete time intervals, this summing (or integrating) process is easier to see. Suppose we put a new brand of marmalade on the shelf in our grocery store and it follows this pattern of dying (being purchased or broken or stolen):

Day	Number dying (N')	Total dead ($N_o - N$)
1	1	1
2	3	4
3	7	11
4	16	27
5	8	35
6	4	39
7	1	40

The relationship between the second and third columns is clear—the total dead is merely the sum of all the daily deaths.

This is precisely the same relationship we see between the N' curve of Figure 3.8 and the N curve of Figure 3.9. Each shows the same death rate—slow at first, then more rapid, peaking at M, staying high a little longer, then levelling off at a very low rate. N' accentuates the *rate*, but N focuses on the overall consequences of that rate maintained over time for the actual *size* of the aggregate. The N curve for the marmalade is shown in Figure 3.10.

The location and value of M does not actually give a complete description of the aggregate behavior. If the N' curve actually is a normal distribution, with a mean value of M, then a second number will provide all the additional information necessary to draw the ogive exactly. This simplicity, of course, is why mathematicians are so fond of the normal distribution, and thus have a tendency to believe the world is more "normal" than even the most casual inspection would discover.

This second number characterizes the *width* or *spread* of the single peak. Mathematicians use either the "variance" or the "standard deviation," implying precise mathematical properties. But any number which

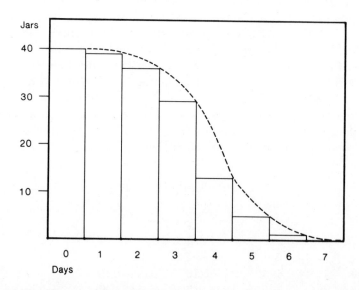

Figure 3.10. The ogive derived from the pattern of disappearing marmalade given in tables in the text resembles the ogive of Figure 3.9, though derived from a small number of items.

characterizes the spread of the peak gives us a somewhat better feel for the ogive in question. Even a simple verbal description can be used, and we often speak of "fat" distributions and "skinny" distributions. Skinny distributions are found where the variance is small—in this case the mathematical term coincides with our intuition from natural language. All members of the aggregate are quite similar—they show little variance. Once they start dying, there is a *sharp drop* in the N curve's ogive. When the members are dissimilar—the distribution is fat, or has large variance—the ogive flattens out lazily, as we would expect from a fat distribution. Indeed, as the distribution gets fatter and fatter, the ogive slumps down to look more and more like an exponential decay.

With the ogive and the exponential, we now have curves to represent two different kinds of birth-free aggregates, different in respect to their characteristic patterns of depletion. The exponential describes an aggregate whose members are completely *independent* of one another, so that the aggregate depletes faster when there are more of them. The ogive, on the other hand, represents an aggregate whose members are more or less *similar,* characterized by a mean or modal time of death. Its rate of depletion is not dependent on *size,* but simply on the *age* of its members.

Composite Curves

> But as all severall soules containe
> Mixture of things, they know not what,
> Love, these mixt soules, doth mixe again,
> And makes both one, each this and that.
>
> John Donne
> *The Extasie*

The ogive, then, represents an aggregate with a "typical" member, with some variation. All oak trees in a forest are pretty much alike, but they do not generally all drop their leaves on the same day. But what about the maple trees, or the ash?

All members of an aggregate need not be the same type. Indeed, pure aggregates are rather the exception in nature. One of the important uses of death curves is to detect the presence of two or more types in the same aggregate and, perhaps, to identify them.

For example, in designing undersea communications cables using vacuum tubes, much effort is expended to find the most reliable possible vacuum tubes to be used in the relay points, for once the cable is submerged, it cannot be brought up to replace a burned-out tube. Now suppose that in the testing of a big batch of tubes for lifetimes there were two types of tubes mixed together. Perhaps one batch had been processed on a certain machine and had a much longer half-life than the ordinary tubes. In Figures 3.11a and 3.11b we see how the lifetimes of these extraordinary tubes (b) might compare on a logarithmic plot with the ordinary tubes (a).

The type a tubes start with N_0 members and have a half-life of R, while the type b tubes start with M_0 and have a half-life of U. M_0 is only ⅛ of N_0, but U is five times R. Therefore, at the beginning we only see the decay of type a, but after a while we only see the decay of type b. Since the tubes are mixed, the composite aggregate behaves as shown in

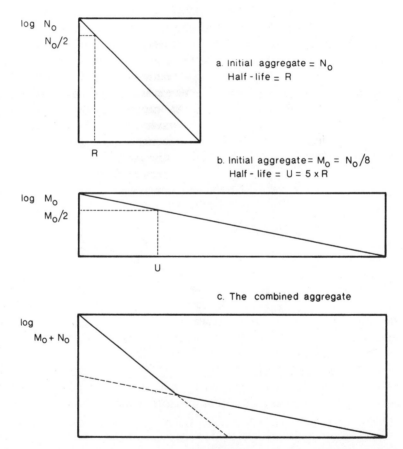

a. Initial aggregate = N_0
Half-life = R

b. Initial aggregate= M_0 = $N_0/8$
Half-life = U = 5 x R

c. The combined aggregate

Figure 3.11.   The existence of two exponentially decaying aggregates mixed into one can sometimes be inferred from the behavior of the combined aggregate, particularly when plotted on a logarithmic scale so that two straight lines can be seen.

Figure 3.11*c*, which on this logarithmic plot is easily seen to separate into two straight lines. Thus, by observing only the aggregate behavior, *c*, we can infer the presence of two otherwise indistinguishable types, *a* and *b*.

Something similar can be done with ogival decay curves, especially if the peaks are narrow and well separated. For instance, there may be two types of oak in our forest, with the dropping of leaves distributed as in Figure 3.12*a*. If we cannot distinguish these trees in any other way, we will see a composite leaf fall as shown in Figure 3.12*b*. We may infer the

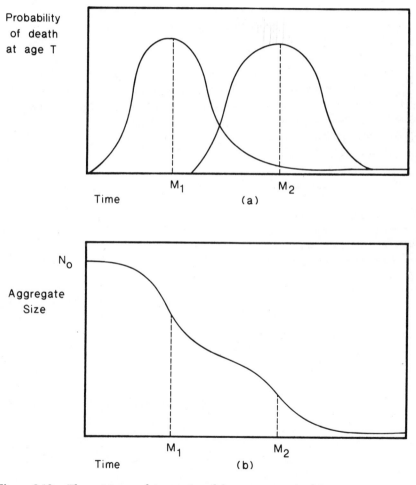

Figure 3.12. The existence of two uni-modal aggregates mixed into one can sometimes be inferred from the behavior of the combined aggregate, which will deviate from a simple ogive.

presence of two "types" by the two steep places, M_1 and M_2—as long as they are not too close together. We would, of course, have similar trouble distinguishing two exponentially decaying aggregates mixed together if their half-lives were quite similar.

Reasoning in this way from black box information on aggregate behavior may be quite useful, but several cautions are in order. First of all, by the Law of Indeterminability, we do not know for sure whether there are two types in the aggregate or whether there are two different environmental conditions. Moreover, as we try to extend the method to more and more types, the composite curve begins to get more and more ambiguous. Even in Figure 3.12b we begin to see how a composite of s-curves might begin to look like an exponential.

Nature does not usually supply us with such nice curves, either. Instead, we might have an assortment of data points, as shown in Figure 3.13. In this Figure, the data "points" don't seem to be points at all, but are represented as crosses. No empirical observation is absolutely accurate, although our habit of representing observations as points might give that impression. Instead, the point represents our best guess as to where the observation actually lay, and using a cross instead of a point permits us to indicate how far the observation might be in error in either direction.

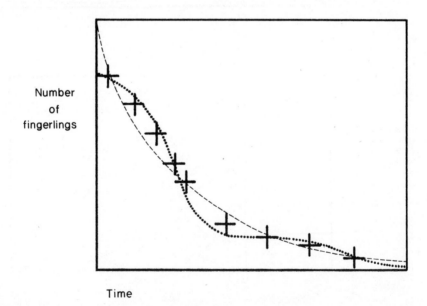

Figure 3.13. Which curve best represents the data points? When there is uncertainty in the data, we may be unable to separate two or more mixed aggregates, no matter what method we use.

In one of the conventional usages, if we draw an ellipse around the cross, it will contain the "true" data point two-thirds—or sometimes 95 percent—of the time.

Suppose that the data points—or data ellipses—in Figure 3.13 represented observations on the number of fingerlings in an enclosed experimental pond. The methods of counting fish in a pond are rather uncertain, which accounts for the large range of potential error in estimating the total number of fingerlings remaining from a single hatch after a given amount of time. The fish breeder, in examining these curves, is trying to determine if some exponential removal process is at work (shown by the broken line), or whether two distinct types of fingerlings make up the school, each dying according to an ogival depletion (shown by the dotted line).

Given the uncertainty of the data ellipses, what is the breeder to conclude? Which of the two curves "best" represents the "true" situation? Is some exponential process removing the fish? Or are there two types of fingerlings present? To get an answer, the breeder would have to turn away from general systems theorizing and get into the pond. General systems considerations can suggest which kinds of depletion curves to try, and even which additional data points might play a critical role in distinguishing between the two curves. But no amount of general systems thinking is going to count the fish. Though general systems thinking may not be entirely scientific, it's not magic, either.

QUESTIONS FOR FURTHER RESEARCH

1. *Antique Cars*

Go to an auto graveyard (it doesn't have to be by the full moon) and count cars of various makes and model years. What can you infer about the original car populations from these junked populations? Will data on yearly production of models help you in your inferences? Can you get some of the same information by seeing which cars are in the bottom layers?

2. *Agricultural Genetics*

One of the objectives of the breeding of domesticated plants is to obtain uniformity of one sort or another. For example, good popping corn will not have too great a variation in popping time, otherwise many kernels would burn, while many others would remain unpopped. On the other hand, the kernels must not be too uniform, otherwise we would not have popping corn, but exploding corn.

Other variables for which uniformity is commonly sought are date of ripening, size of fruit, color, and height of fruit or grain on plant. Discuss the advantages accruing from various uniformities and the consequences of pushing the uniformities beyond certain limits.

3. *Statistics*

The vast body of statistical literature is concerned with the study of "types" or "central tendencies." But it is also possible to study the statistics of extreme cases. Discuss situations in which we are more interested in extreme cases than typical cases, and how we might go about studying them, both mathematically and observationally. (Reference: Gumbel 1958)

4. *Education*

Some educators say that certain ethnic groups do poorly in school because of their early training. Others say that their behavior is genetically determined. What kinds of observations need to be made to determine which faction is correct? Can we rule out a third possibility—that somehow the behavior of all the children in one ethnic group is somehow connected? In particular, what is the possibility that their behavior is connected through the minds of the educators of either persuasion who classify them as an "ethnic group"?

4

Reasoning
About Aggregates

Have you seene but a bright Lillie grow,
    Before rude hands have touch'd it?
Ha' you mark'd but the fall o' the Snow
    Before the soyle hath smutch'd it?
Ha' you felt the wooll o' the Bever?
    Or Swans Downe ever?
Or have smelt o' the bud o' the Brier?
    Or the Nard in the fire?
Or have tasted the bag of the Bee?
O so white! O so soft! O so sweet is she!

        Ben Jonson
        A Celebration of Charis in
        Ten Lyrick Peeces,
        4. Her Triumph

Like the poet, blinded to the blemishes of the one true love, we think about aggregates as if they were as pure as the "unsmutch'd Snow." At first, all we see is the purity, and it is the purity that allows us to think about aggregates at all. But, like the lover, we soon must face the rude realities of the world with all its blemishes. Then thoughts of purity may hinder the maturation of our reasoning, just as they hinder the maturation of love.

Cooperation and Competition—The Law of Collapse

With friends like you, who needs enemies?
        New York expression

We could elaborate almost endlessly on ways in which the distribution of ages at death determines the behavior of an aggregate, a subject well covered by "reliability theory" (Barlow and Proschan 1965). Our purpose, however, is not to develop a mathematical theory of redundancy but merely to indicate the kind of reasoning that can be done with depletion curves. Therefore, we turn now to a consideration of aggregates in which the survival of one individual is in some way dependent on the survival of others. A convenient classification of these systems is into those where a *bigger* aggregate enchances the individual's survival (safety in numbers) and those where a *smaller* aggregate enchances individual survival (everybody for themself). In other words, the individuals either *cooperate* or *compete*. Clearly, the distributions we have just studied—where the individuals act independently—lie exactly on the boundary between cooperation and competition.

In a purely cooperating aggregate the initial performance, when the population is still large, will be superior to an otherwise similar competing population because there are lots of members to help one another. As the population decreases, however, the advantage gained through cooperation diminishes faster and faster, until a "collapse" occurs, as shown in the four curves plotted in Figure 4.1. The curves ending at A, B, C, and D, respectively (notice how a definite end time can be predicted in these cases) show progressively stronger relationships between the survival times of the members. In fact, the limiting case has already been shown in Figure 3.2 as the performance of the serial circuit.

Good examples of pure cooperation are the erosion of a solid aggregate at its surface, a plane aggregate along its sides, or a linear aggregate at its ends—sand grains in a rock worn by water, bacteria on a plate surrounded by penicillin, or hydrocarbon molecules in a candle burning

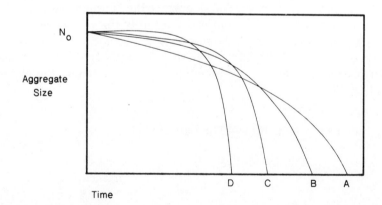

Figure 4.1. The depletion curves for four "cooperating" aggregates, with A showing the *least* cooperation and D showing the *most*—and therefore the sharpest collapse.

at both ends. In each case, individuals at the *center* cannot be exterminated before their neighbors closer to the *boundary*. Since the volume of a solid diminishes relatively faster than its surface, the ratio of surface to volume grows as the aggregate shrinks, and the depletion accelerates.

Another large class of cooperative aggregates does not have this spatial relationship, but has a load-sharing property along with a load-dependent lifetime for the individuals. Suppose, for example, our parallel circuit of Figure 3.1 were built so that when one light failed, the illumination was kept constant by making each other bulb 20 percent brighter. Such an added load on the remaining bulbs could shorten their lifetimes. On this basis, we can easily see the relationship between Figure 3.2*b* and Figure 4.1. If the lifetimes of the remaining bulbs are shortened by the increased load, the space between successive *T*'s in Figure 3.2*b* will, on the average, decrease, yielding, for large aggregates, curves like those shown in Figure 4.1. Such behavior may be seen in the collapse of structures, the breaking of braided cords or wires, or in the dissolution of organizations as members start to leave and put a bigger share of work on the remaining members.

Lest we leave the impression of negativism by these examples of collapse, breaking, and dissolution, remember that *remaining in the aggregate* might be considered "bad," so that the extinction is "good." The spread of certain kinds of knowledge might follow the curves in Figure 4.1 with the aggregate being the unknowing ones, who are subject to increasing pressure to learn as the knowledge spreads. Alternatively, the curves might represent the number of people who have not yet acquired the latest fad of technology, a telephone, an inside toilet, or a pop record.

In these examples, the aggregate may be kept from extinction by a change in the forces at work when the aggregate gets small, and rarity itself becomes a desideratum. Such a change would add a tail to the bottom of the curves, converting them to ogives with no definite end points.

Suppose the action of increasing value with decreasing numbers has been in operation from the beginning. What would be the aggregate behavior? If the desire to preserve the rare or unique were accompanied by an equal desire to destroy the commonplace, the initial movement of this competitive aggregate would be toward extinction at a more rapid pace than the exponential, where the survival of each was independent of the existence of others. At some point, however, the conserving force of rarity would begin to dominate, and individuals surviving to that time would tend to last longer than their independent counterparts on the exponential decay curve. This can be seen in Figure 4.2, where two such curves are compared. Having survived the "competition" from their shorter-lived siblings, these survivors have a much easier time in their old

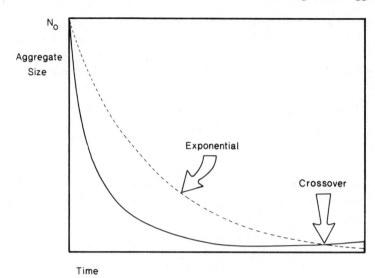

Figure 4.2. Depletion of a "competing" aggregate is compared with the depletion by exponential decay of an "independent" aggregate. The competing aggregate initially does worse than the exponential, but in the long run has a few survivors which offer each other little "competition."

age. In fact, if members die only because of *other* members, the last members will *never* die, since there will then be no others to cause its death. In that case, the curve in Figure 4.2 would never reach zero. Naturally, we know that nothing survives forever, though the elaborate care given to the preservation of rarities such as coins, paintings, or rich celebrities, often makes this an appealing approximation.

Note also that it would be extremely difficult to distinguish the "competitive" curve from the composition of two exponentials, such as we saw in Figure 3.11. It would be easy to imagine that instead of competition we had a mixed aggregate of "good and bad survivors."

But if the amount of competition is fixed once and for all, and only depends on the size of the aggregate, the fiercest competition will always be at the beginning, when the aggregate is largest. If the competition is stronger for one aggregate than another, the initial portion of its curve points down more sharply. In the most competitive aggregate imaginable, the curve becomes completely *L* shaped.

But similarly, the more *cooperative* the members of an aggregate are, as shown in Figure 4.1, the more the aggregate curve approaches an *L*, only turned upside down. In other words,

Any strong association among lifetimes—be it cooperative or competitive—leads to sudden depletion.

This is the *Law of Collapse.*

The Law of Collapse limits the way in which we can reason backward from the black box observation of a collapsing aggregate. We could be seeing the beginning of jealous competition, or the end of devoted cooperation. Once our world begins to tumble around us, departed friends are as bad as determined enemies. When the ship goes down, what matter if we drown for lack of a helping hand or because others cannot swim and drag us under?

The Law of Typology

There is one important thing that [chemistry and physics] have in common and that distinguishes them from biology: they are inherently and legitimately statistical sciences, dealing with enormous numbers of identical components. If we have a bar of pure gold . . . or a solution of pure glucose, we can be sure that every gold atom, every glucose molecule, will behave in entirely the same manner. Should we find anomalies, we would not speak of physical variation or chemical variation; we should call it contamination. But we have all heard of biological variation. (Chargaff 1969)

The phenomenon of sharp collapse was also seen when we drew ogives from life tables. The narrower the peak—the smaller the variation from the mean—the sharper the decline in the depletion curve. Similarly, the broader the peak—the greater the variation—the more gradual the decline.

We may take the point of view of starting with a distribution of independent individuals or we may take the view of dependent individuals and try to measure cooperation or competition. But in looking at the depletion curve only, we cannot tell which is the "true" structure, because there is for any curve a corresponding model of each type. Indeed, there are any number of mixed models, and as we know from the Law of Indeterminability, there are also models in which the curve results from the influence of an unperceived environment.

The difficulty in pinning down the "source" of an aggregate's behavior is not merely academic. Sometimes, the individuals of the aggregate are themselves aggregates, so an appeal to "intrinsic behavior" only pushes the problem one level deeper. The same difficulties remain at the new level, for pushing the problem down to a scale below human observation does not give much psychological relief. In our society we are trained at age four not to carry a series of "whys" more than two or three levels deep.

To illustrate another difficulty, suppose we observe that industrial strikes tend to follow a curve like Figure 4.2, if we plot the length of each strike versus the number of strikes of that length over a large sample of strikes. Now, all the strikes in a given industry, or in a given year, form an aggregate which we feel intuitively is not "connected." Thus any behavior of this aggregate of strikes must come from an intrinsic character of strikes as individual entities, or from some characteristic of the "environment" in which that strike "lives." Environment, in this case, can be characterized by the attitudes of the adversaries in the strike, the pressures of the community, or the intervention of government.

The "intrinsic character" point of view says that some strikes are from the beginning inherently more difficult to settle than others. The "environmental" viewpoint says that as a strike drags on, the adversaries get more and more embittered by the strike itself, making the grievances more intractable. Which is right—the one that attributes behavior to initial conditions, or the one that invokes inputs over time? Only investigation beneath the level of the black box could possibly resolve the issue. Moreover, the one possibility we discarded so casually could also turn out to be correct: the strikes *are* connected, in the sense that labor and management may measure their own strike against their observation of the kind of settlements others have made.

However such an investigation should turn out, its progress would be guided by the investigator's knowledge of aggregate process as well as types of underlying individual processes. Studying individuals by studying aggregates is a favorite method of both science and art. Although it is quite powerful, the method is frequently misapplied, yielding the fallacy of "typological thinking." Typological thinking results from the unquestioned assumption that the behavior of an aggregate is merely the sum of individual behaviors, each of which has essentially the same properties as each of the others. This is a case of the *Decomposition Fallacy* applied to aggregates.

In its crudest form, typological thinking imagines that there is some reality to the "average" individual. For example, suppose we surveyed the audience attending the ballet at a Saturday matinee. It happens to be sold out to the local Masonic temple whose members are taking their young daughters for a bit of culture. In such an audience, we might have 5-year-old girls about 3 feet tall and weighing 50 pounds, plus an equal number of 35-year-old men about 6 feet tall and weighing 200 pounds.

From such a survey, we could easily conclude that the "typical" aficionado of the ballet was a 20-year-old fat midget, four-feet-six tall and weighing 125 pounds—and a hermaphrodite, to boot.

While this example may seem far-fetched, the typological error is all the more dangerous when it is more subtle. Whenever we average both weights and heights for a nonuniform group, the "average" individual will seem to be "overweight," since weight is not proportional to height but more closely proportional to the cube of height. Thus, if we had the following three men:

Height	Weight
6′ 2″	219
6′	200
5′ 10″	184

their "typical" member is 6 feet tall and weighs 201 pounds, if we take a simple average of both height and weight. A more meaningful average would be about 200 pounds. That one extra pound may seem trivial, but to us fatties who have to live with published tables of "typical" weights, it means a lot of forgone desserts.

To the typologist, then, behavior of aggregates is seen as behavior of more or less identical individuals. Suppose he works in a room illuminated by a vast aggregate of light bulbs. These burn out one by one as the days go by. When 10 percent have failed, he imagines that all the bulbs are still working, but that each has been dimmed by 10 percent. Or, if the entire array goes out before his eyes, he does not speak of co-operation or competition, but only shrugs and says it was preordained. Light bulbs, he sighs, have their natural span of years, at which time they all naturally die.

The fallacy in typological thinking, as with all illusions, lies not in the explanations themselves, but in the failure to *consider* other explanations. Particularly important is the failure to consider possible diversity in or interaction among individuals—either of which might lead to the same observed aggregate behavior as some collection of uniform members.

Plato, with his philosophy of *eidos*, was probably the typical typologist, if we dare use that expression. To Plato, any variation observed was an unreal or irrelevant deviation from the ideal type, which was truth. Until recently, Plato's thoughts on this matter dominated scientific thinking. Then, the biologists began to reverse the priorities. In modern biology, as Chargaff implies, it is usually the variation that is real and the mean value which is the abstraction. This lesson from biology gives a worthy example for, say, the social sciences to follow, though a bit of caution is in order. Typologists could be right when they claim that a particular aggregate consists of preordained types. Such aggregates do indeed exist. Where they go wrong is in believing they can make that conclusion solely from black box observation of the aggregate—but a "variationist" could be wrong, too, seeing only the black box.

Because the error in reasoning is nowadays most often done in the direction of typology, we shall call the general principle the *Law of Typology*, by analogy with the *Law of Indeterminability*. For like that law, it tells us something about the limitations of what the observer can do:

> *We cannot with certainty attribute observed constraint to either a type of individual or a kind of interaction.*

We could have called this principle the Law of Variability, or even the Law of Interdependence. Like many of our others laws, it is a law of observer rigidity. In school—in mathematics courses, for example—we learn *how* to derive the behavior of the typical individual from the behavior of the aggregate. What we don't learn is *when* to derive it. While mathematics gives us the "know–how," laws such as the Law of Typology try to give us a little "know-when." If the Law of Typology flashes into our mind at precisely the right moment, it may lead us to consider alternatives before we waste too much time on the calculation of averages.

From Deaths to Births

Geophysicists are understandably excited, because they have clearly detected continental drift. This finding comes as no surprise to the public official, who has already observed that the island of Jamaica, a huge exporter of bauxite, is gradually drifting—in the form of a unicellular layer of aluminum beer cans— onto the United States and covering us. (Savas 1971)

Because the strategy of aggregation is pervasive, one of the most *general thinking* strategies is to ascertain what we can about the ebb and flow of aggregates. To form a correct *general* notion, we seek an intuitive feeling for the variation in the aggregate's size over a period of time. As we have learned, this thinking strategy sometimes employs variations of a simple equation and sometimes employs sketches of N versus time, $\log N$ versus time, N' versus time, or perhaps some other transformation that highlights what we want to know about the ebb and flow of N.

Up until now, however, we haven't really considered the ebb and flow, but only the ebb. We have consciously restricted our thinking to aggregates in which B, the birth rate, equals zero. But much of our thinking about death processes—not all, of course—can be carried over rather simply to help us think about processes involving births.

Before looking at any explicit birth processes, however, we must observe the caution that whenever new members may be added to an aggregate after initial creation, typological thinking utterly fails. Consider a restaurant and the number of people (N) sitting in it at any time throughout the day. The value of N depends, of course, on the individual behavior—how long each person takes to eat—and on the effects of interaction—how the service, for example, depends on crowding.

But N also depends on how the *arrival* of new patrons is distributed throughout the day. That is, N depends not only on the initial state, but also on the *input*.

Notice how we might say, in ordinary speech,

"There are *people* in the restaurant from 6 A.M. to midnight."

From this ambiguous statement, we cannot necessarily conclude that

"*Someone*, perhaps Edgar Millefeuille, is in the restaurant from 6 A.M. to midnight."

It might be true, of course, but very likely isn't. Even Edgar, the chef, might have gone out for dinner to a really *nice* place, and thus not have been in his restaurant for all of the 18 hours consecutively.

We have a tendency to make such typological conclusions, because we are inclined to think of systems as *closed to input*. If there were no people entering the restaurant, and if there were people in the restaurant from 6 A.M. to midnight then it *would* follow that *some* individual had to be in there the whole time.

Our typological tendencies are encouraged by a peculiarity of the English language, which sometimes allows us to interpret the word "people" in the first sentence as meaning "specific individuals," as in:

There are specific individuals in the restaurant from 6 A.M. to midnight.

In making this innocent grammatical transformation, we slip into an egregious typological sin.

Even when we feel we have mastered our tendency to typology, we often backslide when tricked by a leading question. Horace Davenport has written an article with the intriguing title, "Why the Stomach Does Not

Digest Itself." As it turns out, one of the answers to this question is that the stomach *does* digest itself, but

. . . the mucosa has a second safeguard against self-digestion in addition to the normal barrier to acid invasion. The mature cells of the stomach mucosa are continuously desquamating, or flaking off, from the mucosal surface and are replaced by new cells produced within the tissue. Indeed, the human stomach normally sheds about half a million cells per minute. Thus, the surface lining of the stomach is completely renewed every three days. (1972:91)

Fortunately for us, we don't have to call time-out 30,000,000 times an hour to tell some stomach cell to "flake off"—otherwise we should surely have ulcers. Instead, we save our sanity and stomach by being typological and speaking of "the lining of the stomach"—as if it were some fixed aggregate staunchly resisting the assaults of our gluttony.

Stomach cells are, of course, generated from other stomach cells through cell division, forming, in our terms, a *population* rather than the more general *aggregate*. But it is not yet time for us to explore the powers of populations, for we have not yet exhausted the possibilities of simpler aggregates. New members need not come from self-reproduction of present members. Most often, the influx of new members can be traced to another aggregate, like the great waves of Irish migration to America during the potato famine, dandelion seeds floating onto our perfect lawn from neighbor's unkempt grass, and general systems ideas similarly floating from our minds to yours through the medium of this book.

At times, what enters one system can be said to "leave" another, but not always. An American citizen who becomes naturalized as a Swiss citizen is dropped from the aggregate of Americans and added to the aggregate of Swiss. But a Swiss citizen naturalized as an Amercian citizen retains Swiss citizenship and is thus counted in both aggregates. Or, to cite another case, when a professor bestows some knowledge on a student, that knowledge also remains with the teacher, though some professors act as if knowledge is like a book—lost if lent to a student.

Where there is proper partitioning, the act of "dying" for the parent is the same as the act of "being born" for the child—or "daughter," as it is conventionally called. Nobody knows why we don't call an atom of lead the "son" of an atom of radium, but whether we call them daughters or sons, these children follow a birth curve that is the upside-down of the death curve of their parents.

In some circumstances, the entire growth history of a child aggregate can be derived from the death history of its parent aggregate. If $D = 0$ for the child, what goes out of the parent goes into, and stays in, the child. The child's accumulated births are the parent's accumulated deaths. In

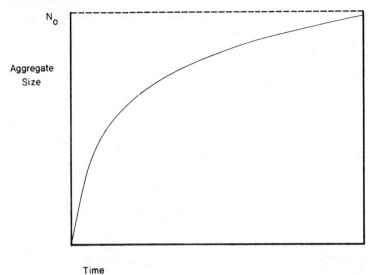

Figure 4.3. The growth curve of a "daughter" aggregate whose parent is the exponential decay of Figure 3.4.

Figure 4.3 we see the growth curve of the daughter of the exponentially decaying parent of Figure 3.4. In Figure 4.4 we see the growth of "Son of Ogive," whose father's demise was characterized by Figure 3.9. Finally, in Figure 4.5, we see the rapid ascendance of the daughters of the poor, collapsing parents of Figure 4.1.

These curves all show that the *Principle of Indifference* can be applied to birth and death. By the Principle of Difference, we know that by turning a problem upside down—or, more precisely, by turning the observer

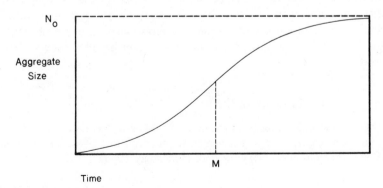

Figure 4.4. The growth curve of a "daughter" whose parent's death curve was the ogive of Figure 3.9.

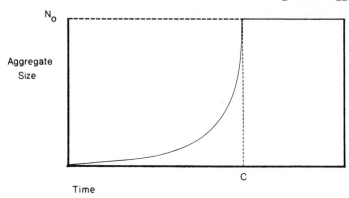

Figure 4.5. The growth curve of a "daughter" whose parent's death curve was the collapsing curve, C, of Figure 4.1.

upside down—we can often bathe an obscurity in a fresh light. Physicists, at least, have known this technique for years. Instead of tracing electrons all the time, they sometimes trace "holes," or places where "electrons ain't."

Social scientists are also beginning to discover this inversion technique. Harrison White (1970) for instance, introduces the concept under the name of "vacancy chain." He uses it to study the mobility of job holders by studying the way vacancies are pushed down a hierarchy of fixed jobs.

Of course, the ancient Chinese knew this technique, too. As the story goes, a poor monk was walking behind some other poor monk, thinking to himself, "I must surely be the poorest monk in the world, for here I am forced to eat the pieces of food that another monk finds unfit, and drops on the path behind him." As he stooped to obtain his next morsel, he chanced to notice a third monk, behind him, picking up *his* leavings. The general systems moral is clear: if one system's deaths are another system's births, nobody can be sure of having the correct point of view—even the poorest monk.

Reasoning in the Presence of Births

One of the first objects of an inquirer, who wishes to form a correct notion of the state of a community at a given time, must be to ascertain of how many persons that community then consisted. (Macaulay 1968)

If you would like to gain an appreciation for the difficulties of social sciences in relation to the easy sciences like physics or chemistry, you might start by trying to answer Macaulay's question about some community with which you are familiar. You need not make the problem triply difficult by asking about some time in the past. Simply answer the question, "How many people live today in the town of Watsonville, or the Russian Quarter neighborhood, or the Butterball House apartment building?"

Among the many difficulties you are sure to encounter, the problem of people moving across the boundary you set is perhaps the most difficult. For that reason alone, we can understand the predilection that anthropologists and sociologists have for studying *isolated* communities. They can reasonably well handle people leaving, but they really worry, just like the peasants themselves, when someone new arrives on the scene. And, of course, if people can return, then how do you even know when someone has left? We know of many cases where villagers returned as if from an overnight outing when they had in fact been gone for fifty years or more.

When choosing a site for study, using available statistics, it's even difficult to tell whether or not $B = 0$. The most certain indication that we're not dealing with a $B = 0$ aggregate is that we see an increase in N at some point in the statistics. Clearly, if B were zero, the aggregate could never increase in size, since N' is never positive, which would indicate an upward slope.

Of course, just because we never see an increase in N, we dare not conclude that B is always zero. If B and D are matched in some way so that they cancel each other over the period of our observation, we may not observe an increase, even though there are birth processes taking place. For example, the number of taxi medallions in New York City is fixed, but drivers come and go.

Usually, however, with refined enough observation over a long enough time we will detect even a tiny increase. Any increase, no matter how small, will tell us that new members are being introduced, though we may have no clue about their source. When Pandora opened her box, or, rather, opened the world to the miseries in her box, she had nothing on the systems thinker whose system is found to be open to the input of new members. Once new members can be introduced, the questions become much more difficult, and some completely new questions make their first appearance.

Beyond the question of stability, for instance, lies the whole question of identity. For birth-free systems, we could usually be content to identify a system by size alone. If we're still satisfied with this simple method of identification, then neither B nor D will concern us as long as $B = D$. Suppose, for example, that a tiny crack in our screened porch admits fresh flies at precisely the same rate we are killing old flies. To us, it appears that the fly aggregate is remaining exactly the same, in spite of all our most virulent efforts at eradication.

But suppose the newcomers arriving through the crack are wasps. We now have a different problem on our hands, though we may not recognize it immediately. Eventually, we may be obliged to consider that the system is a new and different one, or take the consequences.

In short, as soon as we can observe variation in the *composition* of an aggregate, and not merely its *size*, it becomes dangerous to remain typologists.

This dilemma has been of immediate concern to students of rural populations. In the United States, newspapers bemoan the decline of the family farm and the ascendance of agribusiness. In Western Europe, and particularly in France, where Marx described peasants as "a lot of potatoes huddled in a sack," social scientists and government officials fret over the rapid rate of rural exodus—the emigration of those "potatoes" from rural communities to urban centers. In many cases, however, following the undifferentiated view that sufficed for Marx's reasoning in the nineteenth century leads to scientific and administrative nonsense.

Consider the case of the French Alpine commune of St.-Colomban-des-Villards, which has dropped from 1061 persons in 1911 to 252 in 1962 (Bozon 1970:217). This commune traditionally supplemented its agro-pastoral subsistence base by sending men into the lowlands, as far off as Grenoble and Marseilles, for winter employment. The traditional winter occupations were all itinerant: sweeping chimneys, carding hemp, and selling dry goods. Men and their sons came down from the mountains after the fall harvest, gradually worked their way down to the cities, and then slowly reascended for spring planting.

We see immediately, then, that these peasants, at least, were not an undifferentiated sack of potatoes, and were *already* making a rural exodus *every year*. Modern technology put the sweeps and carders out of business, so the first effect was to make them move from the cities to spend more time in the country. Presumably, this sudden increase in the population made it impossible for all to survive on the agricultural production. Some were eventually driven to settle in the cities.

The fate of the itinerant merchants was different, though in the end they, too, were counted in the rural exodus. Those merchants who were able to accumulate sufficient savings invested them in more profitable and reliable permanent shops, and thus stopped being peasants altogether.

Even after we've discovered this much structure, we still haven't exhausted the problems of accurately describing the rural exodus. We know that the population always fluctuated throughout the seasons, so we could be seeing an effect caused simply by a different time scale. For birth-free aggregates, we would see only declines or constancies, no matter when we made our observations. Here, as people flow in and out of the village, there could be trouble. For instance, in 1911, there was little chance that a sociologist would visit an isolated commune in the Alps during the winter when the population was smallest. By 1962, however, skiing was a well-developed pastime enjoyed even by sociologists, and many grants were obtained for winter studies of Alpine communities. What looked like rural depopulation could easily have been a consequence of *summer* observation in 1911 and *winter* observation half a century later. In this case, it wasn't, but worse conclusions have been published by researchers who failed to take differentiation of an aggregate into account.

In the case where *B* happens to equal *D*, the temptation to erroneous conclusions is even harder to resist. The French commune of Roussillon (Wylie 1963), for instance, has remained stable in *size*, but not in the *composition* of its population. In both 1946 and 1954, the census tallied approximately 700 people. One could easily conclude that Roussillon wasn't suffering from the rural exodus at all.

Looking in more detail, however, we would find that only 275 of the 700 people were listed in *both* censuses, only eight years apart! Thus 425 names changed between 1946 and 1954, so that the total number of *different* people who lived in Roussillon at some time in that period was at least $275 + 425 + 425 = 1125$. And considering the rapid rate of moving out and in, it's quite likely that there were many who moved in after 1946 and out before 1954, thus missing both censuses and making the actual number of sometime inhabitants larger than 1125. Wylie himself estimates that "the total may well be nearer 2500" (1964:217). Hardly the image of a "stable population."

This "illusion of demographic stability" depended on both the stability of total population (probably determined by the number of available living spaces) and on the existence of a small core population of 275 who were property owners and political powers in the commune. In addition,

. . . the people who move into Roussillon are very much like the people who leave . . . families of about the same composition and professions. (Wylie 1963: 236)

A black box view of Roussillon indicates at least one commune in France that's not succumbing to the rural exodus. But below the black box view, we find that we're really dealing with two different aggregates, one that remains at 275 and the other that begins and ends at 425 but fluctuates in between. Is this second aggregate one system, or something else? With nothing but rough birth and death figures, we've gone as far as we dare.

The Second Aggregate Law

According to the biologist, a physicist cannot tell the difference between a room containing a man and a woman and a room containing two men.

According to the anthropologist, a biologist cannot tell the difference between a room containing a husband and wife and a room containing a mother and son.

According to the psychologist, an anthropologist cannot tell the difference between a room containing Jocasta and Oedipus and a room containing Victoria and Albert Edward.

In general, then, it is not possible to reason backward from the behavior of an aggregate to the behavior of its members, unless we descend, at some point, to the level of examining individual members. Why, then, do people persistently make the typological error? One possible reason is the failure to distinguish between the appearance of an aggregate system from the inside and from the outside, that is, from the points of view of systems *design* and systems *observation*.

Consider first the inner reason for having an aggregate in a system. Going back to our checkers example (at the beginning of Chapter 2), suppose we had a new type of checker, neither red nor black, but red on one side and black on the other. In this case, a "complete set" consists simply of 24 checkers, all of the same type, for I can obtain either red or black ones simply by turning them over. But notice what I gain by this change: if I start with 30 of these checkers in my summer house, I can expect the set to survive for seven years, not four. Viewed another way, I can obtain the same four years of protection by having only three extra checkers, not six.

Whenever I can reduce the differentiation among the pieces of the aggregate, I can expect longer survival for the same expenditure, or the same survival for less. For an automobile, one spare tire gives me a good measure of protection against complete breakdown; but for a tractor, which has two different sizes of wheel, I need at least two. In business organizations, if one of the secretaries gets pregnant and leaves, her work is not so specialized that there is any real difficulty in taking up the slack. But, if one of the vice presidents gets pregnant and leaves, the firm may be seriously hindered in its work.

Of course, some secretaries become so specialized that they are no longer interchangeable with others, which tends to defeat the redundancy of a secretarial pool. Large companies try to keep their employees as interchangeable as possible, even at the executive level—which may account for the "grey flannel suit" image. But, if an organization really succeeded in making all its employees identical, the result would be most unhappy: on the same day, every one of the employees would troop to the president's office to resign because of pregnancy—only to find that the president herself was knitting booties!

We need not restrict ourselves to such contrived examples, but may turn to the world's second most interesting topic—money. Banks hold valuables in two ways, in safe deposit boxes and in accounts. In safe deposit boxes, each person's valuables are kept distinct from those of all others. We would not be satisfied to get someone else's Piaget watch in return for our shares of AT&T stock. But with our accounts, the situation is different. If we deposit five $20 bills, we neither desire nor expect to get those same bills back when we withdraw $100. Though we may think of "our account" in the same way we think of "our safe deposit box," the situation is entirely different from the bank's point of view.

Since all clients want their own jewels or stock back, the bank simply keeps them locked up in its vault. Because nobody insists on getting the *same* money back, the bank can lend our money to someone else and pay our withdrawal with somebody else's. Because the money is *indistinguishable,* the bank may operate on a *reserve* basis, keeping only a small fraction of its total deposits on hand in the form of ready cash to meet each day's withdrawals. Not all depositors are going to need their money on the same day, for each depositor has individual financial requirements.

How much reserve the bank needs is determined by how similar the depositors are. Just before payday at the local factory, the reserve requirement may be higher, for people are taking out money to tide them over the last few days. In the extreme case—where there is a run on the bank—all depositors demand their money on the same day. In that case, the bank cannot operate on reserve at all. In the general case, the more similar the depositors' requirements, the higher the reserve needed, which explains why banks attempt to diversify their clientele.

In general, then, though we want our aggregate members to function in an identical and interchangeable manner *inside* the system, their behavior with respect to passing *out of the aggregate* should be as variegated as possible. Because, in thinking about individual members, the typologists imagine themselves "inside" the system, they stumble into the view that all members should be "the same."

Now we can see what really "distinguishes [physics and chemistry] from biology." There is nothing "inherent" or "legitimate" in their use of statistics—that is, in their use of means or of types. In nature, we very rarely find a bar of pure gold or a solution of pure glucose. In fact, we very rarely find "pure" undifferentiated anything, since purity is one of those observer's fictions. The physicist has to work very, very hard to get her "pure" gold, and even then she will have a mixture of isotopes, unless she works very much harder.

The reason we don't speak of "physical variation" is that the physicist has chosen not to work with naturally occurring systems, but, rather, with the products of her laboratory. Most of her time in that laboratory is spent *eliminating* the variation and keeping it eliminated. Moreover, she must artificially ensure the survival of each aggregate, for a lump of pure gold, and especially a solution of pure glucose, would not survive long without the active participation of the scientist.

The difference, then, is between the sciences that study nature "as it comes," in all its variability, or, rather, as it has survived, and those that must first "purify" in order to study. Biology, in fact, is split in two on this subject, more or less along the lines of molecular biologist—the purifier—and systematist—the naturist. Psychologists are also split, between those who study purified rats and those who study (relatively) unpurified people. The split in the social arena seems to follow more or less the sociology-anthropology axis. In fact, all the sciences have played some on each side of the fence. This has also been called "mechanism-vitalism" and "reductionism-antireductionism."

Historically, these arguments have generated more heat than light, which would indicate that there isn't much content behind them. From our vantage point, all that's behind them is the *Second Aggregate Law*. It states:

> *For aggregate success, members must be the same to the system and different to the environment.*

Phrased this way, the Second Aggregate Law is a paradox that acts as a limit to the strategy of aggregation.

Phrased another way, it tells us about a paradox that acts as a limit to the scientific strategies of reductionism and antireductionism:

> *To see aggregate survival, an observer must ignore differences among members, though differences must exist if the aggregate is to survive.*

The reductionist—interested in "law"—emphasizes the sameness; the antireductionist—interested in "life"—emphasizes the difference. If the reductionist goes to the extreme, his laws are about systems that have zero probability of being observed; while if the antireductionist goes to his extreme, his systems also have zero probability of being observed— for if a system is truly unique, our minds cannot deal with it at all. In the first case the system does not exist, while in the second we have no way of observing its existence. We leave it to the philosophers to debate the difference.

QUESTIONS FOR FURTHER RESEARCH

1. *Space Travel*

One of the Apollo moon flights almost ended in disaster when one of the main batteries exploded. Although three batteries had been provided and redundancy was such that any one of them was sufficient for the flight, both of the other batteries were damaged by the explosion, since the three were adjacent in the capsule. Discuss this design in terms of the Second Aggregate Law, and what could be done to improve it.

2. *Political Economics*

. . . the favorite postulate of the Narodnik economists [is] that "Russian peasant economy is in the majority of cases purely natural self-sufficient economy" . . . All one has to do is to take "average" figures which merge the rural bourgeoisie with the rural proletariat—and this postulate can be taken as proved! . . . This is precisely the kind of data the authors [of the above work] confine themselves to when they speak of the "peasantry." They assume that every peasant sows *the very grain* that he consumes, that he sows *all* the kinds of grain that he consumes and that he sows them *exactly in the proportions* that he consumes them. . . . In Narodnik literature one may also come across the following ingenious method of argument: every *separate* form of commercial farming is an 'exception' to agriculture as a whole. Therefore all commercial farming generally should be regarded as an exception, the general rule should be taken to be self-sufficing economy! (Lenin 1966)

Discuss Lenin's accusations against the Narodniks in terms of typological thinking—both on their part and on his. If possible, refer to the original work, for this sample is only one of a great many cases where such arguments arise.

3. *Historical Demography*

T. R. Malthus, in his famous *Essay on Population* (1978), recognized the central importance of age at marriage in determining the birth rate of a population. He was, for example, strongly impressed with the words of a Swiss peasant:

. . . the habit of early marriages might really, he said, be called *le vice du pays;* and he was so strongly impressed with the necessary and unavoidable wretchedness that must result from it that he thought a law ought to be made restricting men from entering into the marriage state before they were forty years of age, and then allowing it only with *des vieilles filles*, who might bear them two or three children instead of six or eight.

Malthus' lead has been taken up by historical demographers, but a slight change has crept into his original thinking. Instead of age at marriage, demographers generally draw their conclusions based on average age at first marriage. Show how, for example, a population in which half the women marry at 20 and half at 30 could have quite different fertility than one in which all the women marry at 25, though both have the same average age at marriage. Discuss the sorts of precautions demographers must take to avoid such a typological fallacy.

4. *Economic Development*

Capital formation is usually considered by economists to be an essential prerequisite to economic development. Explain how capital formation can be taking place in a society even though average saving is zero.

5. *Hydrology*

When there is a sudden storm over the drainage basin of a river, we may consider this as the birth of a certain aggregate of water. The water leaving the basin through the main river may be considered the daughter aggregate of this storm water—at least in part. Explain why the birth profile of this offspring will, in general, be less sharp than that of its storm-born parent—why, in other words, we don't get a flood every time there is a thunderstorm, and why we sometimes do get floods even when there is no single big storm.

6. *Insurance versus Gambling*

Because of the Square Root of N Law, an insurance company that is four times the size of another halves its risk. That is, it will experience larger dollar fluctuations in claims to be paid, but only twice as large as a company one quarter the size. Therefore, the risk per dollar of premium is less. In the limit, a company with one policy is not an insurance company, but just a bookie.

The behavior of this law, of course, depends on independence of the events insured. Speculation is the exact opposite. If you are trying to make a killing, you pile into a single investment, in hopes that it is not independent, and that it will succeed. Thus if an insurance company wanted to have the maximum chance of gaining a certain margin of profit, it would insure only one kind of risk in one area—and then hope the clients didn't collect.

Explain the methods by which insurance companies try to ensure that their risks are independent, so that they don't collapse when all policy holders come to collect at one time.

7. Facts of Life

. . . a decline in infant mortality is almost exactly equivalent to an increase in births. It is true that new persons arrive without new confinements, but the economic significance of a confinement is small compared with that of the labour of suckling, nursing and rearing, the cost of feeding and clothing, the pressure on houseroom, which follow. We might, in fact, speak of the number of children reaching the age of six months or one year as giving the "net birth-rate," which is more vital than the "gross birth-rate" to the population problem as it presents itself to the economic historian. In the early part of our period a stationary, or slightly declining, "gross birth-rate" implies a rising "net birth-rate." It is, therefore, misleading to contrast a falling death-rate with a rising birth-rate as causes of population growth, without pointing out that death may come at any age, but birth usually occurs about the end of the ninth month. If the lives saved had been those of septuagenarians, the contrast would be real. Actually, it is false. (Marshall 1929)

Discuss the above quotation in relation to the ideas presented in this chapter.

8. Fast Food Restaurants

The fastest of the fast food restaurants depend on pre-preparation of the food for most of their speed. Speed gets them business, but quality also gets them business. Or, rather, lack of quality may lose them business. In order for quality to stay high, the food must be prepared not too far in advance. In order for the food to be prepared not too far in advance, business must be brisk, so the food turns over quickly. In order for business to be brisk, the food must be of some reasonable quality. In order for quality . . . But we already said that.

Describe the potential for collapse of a fast food restaurant based on this strong association between quality and turnover. What can the owners do to take advantage of the good side of this phenomenon, while not being hit by a collapse if business temporarily slows down?

5

Modeling Differentiated Aggregates

I sing of *Brooks*, of *Blossomes, Birds,* and *Bowers:*
Of *April, May,* of *June,* and *July*-Flowers.
I sing of *May-poles, Hock-carts, Wassails, Wakes,*
Of *Bride-grooms, Brides,* and of their *Bridall-cakes.*
I write of *Youth,* of *Love,* and have *Accesse*
By these, to sing of cleanly *Wantonnesse.*
I sing of *Dewes,* of *Raines,* and piece by piece
Of *Balme,* of *Oyle,* of *Spice,* and *Amber-Greece.*
I sing of *Times trans-shifting;* and I write
How *Roses* first came *Red,* and *Lilies White.*
I write of *Groves,* of *Twilights,* and I sing
The Court of *Mab,* and of the *Fairie-King.*
I write of *Hell;* I sing (and ever shall)
Of *Heaven,* and hope to have it after all.

Robert Herrick
Hesperides: The Argument of His Book

Paeans to purity, poetical or mathematical, soon lose the attention of a lively reader. The rich differentiation of Herrick's world fascinate us long after Jonson's Lillie has faded in fact and in our memory. The pure, simple aggregates with which we started our mathematical explorations are, simply, inadequate as models of "*Brooks,* of *Blossomes, Birds,* and *Bowers.*" We turn, therefore, to the task of breaking down the undifferentiated world, piece by piece, into more appropriate models of their richness.

The State Vector

Next, if we look more closely at the origins of migrants, we do not discover what both the gross statistics and the stereotypes of the older historiography might lead one to expect, that is to say, an undifferentiated mass movement of "peasants" or indeed "artisans" thronging towards immigrant ports from vaguely conceived "countries of origin" like "Italy," "Germany," or even "Poland" or "Ireland." Even discounting the obvious fact that such labels were, for most of the period, inept labels for geographical expressions or political provinces, the picture is false. Seen through a magnifying glass, this undifferentiated mass surface breaks down into a honeycomb of innumerable particular cells, districts, villages, towns, each with an individual reaction or lack of it to the pull of migration . . . We must talk, not of Wales, but of Portmadoc or Swansea, not of North or South Italy, but of Venetia Giulia, Friuli, Basilicata and Calabria. . . . There were villages in the Peloponnese, Basilicata and Friuli where boys grew up expecting to emigrate: there were sections of New York where immigrants from individual Italian districts occupied separate streets, with often mutual hostility. Only when we examine such districts and townships and trace the fortunes of their native sons, do we begin to understand the true anatomy of migration. (Thistlethwaite 1964)

The general systems thinker will not be surprised to find that histories of mass movements of undifferentiated peasants often turn out to be inaccurate. Nevertheless, for many purposes the view of peasants as so many potatoes in a sack might be more than adequate—regardless of how insulting it might feel to the peasants themselves. It's not a question of "truth" or "falsehood," as Thistlethwaite seems to believe, but of what questions you want answered.

Certainly, there was no need to reach down to the level of the individual peasant village, let alone family, for Marx to develop his sweeping political economy. On the other hand, Wylie's understanding of the rural exodus from Roussillon required analysis far below Thistlethwaite's "particular cells, districts, villages, towns." After all, *districts* don't migrate; *people* do. And if an epidemiologist wants to study the effect of the plague on European peasants, an even finer level of aggregation may have to be examined—the cells in each individual's blood. Yet even this fine discrimination isn't enough for the human geneticist, who may want to examine each individual's aggregate of genes that is contained within any single cell. And if the peasant dies and leaves bones in a future archaeological site, some future digger may want to look at the very carbon atoms—*almost* as fine a discriminated aggregate as we can use.

The problem, then, is not to find the "true anatomy" of anything, but the level of anatomical analysis that will serve our purposes, whatever they might be for the moment. We know already that aggregates underlie almost every system we can imagine, but we have so far confined ourselves to the study of pure aggregates. Though these simple aggregates are interesting in themselves, most of their value is as a stepping stone to the study of more differentiated systems. Armed with our understanding of simple aggregates, we can now undertake the study of the behavior that emerges when several aggregates are brought together and considered as a single system.

One aggregate could be characterized by a single number. A mixture of two or more aggregates, however, will have to be characterized by two or more numbers, one for each element. Mathematicians call such a set of numbers a *vector*, and have a healthy bag of tricks for vector manipulation.

A vector, to those unfamiliar with the notation, can be written out in parentheses, as in $(55, 921, 38, 227)$. This is a vector of four elements, which might represent the number of migrants from (Venetia Giulia, Friuli, Basilicata, Calabria). The overall vector might be given a name, such as Italians.

In general, if there are g elements distinguished in the aggregate, there will be g elements in the vector which describes the *state* of the aggregate at some moment in time, as in

$$(N_1, N_2, \ldots, N_g)$$

Another way of characterizing the state of the aggregate, of course, would be to sum all the elements, $N_1, N_2, \ldots, N_g$, giving a single overall aggregate. Note, however, that there are *many* different vectors that sum to the same value. If we merely treat the differentiated aggregate as a single type, we lose a lot of information contained in the state vector.

The state vector, however, doesn't hold all the information about the aggregate either. All the things we know about single element aggregates can be carried over into differentiated aggregates by simply interpreting the equations as *vector equations*. Wherever N appeared before, we now take it to represent the vector

$$N = (N_1, N_2, \ldots, N_g)$$

In particular, the fundamental equation of any system

$$S_{t+1} = f(S_t, I_t)$$

is the same, with S being a vector. Thus the equation says that the state vector (S) at time $t + 1$ depends on (f) the state vector at time t (S_t) and the inputs (I_t) in the interval from t to $t + 1$. (I_t, of course, could also be a vector.)

Since the state vector of a differentiated aggregate system is N, we can write

$$N_{t+1} = f(N_t, I_t)$$

which actually stands for a set of g equations, one for each member of N. In each of these equations, the value of the left-hand element, in general, can depend on any or all of the elements in the aggregate. Therefore, although this vector equation *looks* just like the previous single equation, it conceals many intriguing possibilities for *relationships* among the elements of the aggregate. Our first job, then, is to learn how to characterize relationships.

Constructing a System of Equations

The physicist starts by naming his variables—x_1, x_2, . . . x_n. The basic equations of the transformation can then always be obtained by the following fundamental method:

1. Take the first variable, x_1, and consider what state it will change to next (called x^*_1).
2. Use what is known about the system, and the laws of physics, to express the value of x^*_1 (i.e. what x_1 *will be*) in terms of the values that x_1, . . . x_n (and any other necessary factors) have *now*. In this way, some equation such as

$$x^*_1 = 2ax_1 - x_3$$

is obtained.
3. Repeat the process for each variable in turn until the whole transformation is written down.

The set of equations so obtained—giving, for each variable in the system, what it will be as a function of the present values of the variables and of any other necessary factors—is the *canonical representation* of the system. *It is a standard form to which all descriptions of a determinate dynamic system may be brought.* (Ashby 1956:35–36, with some adjustment of notation)

We have a number of ways of characterizing relationships among interconnected aggregates, and we can start our discussion with any of them. We know that the chronological graph is *not* particularly good at showing relationships, but it is not the only arrow in our quiver. We may, for example, write down *equations*—a word that chills the heart of the mathematically untrained. Because of this fear of equations, we intend to start our discussion of models of structured aggregates with a discussion of how this structure can be represented in systems of equations. That way, having conquered the phantom we fear most, the other methods will not daunt us. And conquer it we will, for there is no great mystery in equations—as long as we don't have to *solve* them. That, as usual, we shall leave to the computer and to the mathematician.

If the state of a system can be modeled as a vector of g subaggregates, the method of writing down the necessary equations in a general form is rather well known. There must be, first of all, g fundamental aggregate equations in the form

$$N_{t+1} = N_t + B_t - D_t$$

The connections among the equations are through parts involving the other aggregates. What is dying in one is being born into another.

In many cases, where the behavior is sufficiently well mannered, the equations may be differential equations:

$$N' = B - D$$

Because this form suggests a huge tub with the bibcock (B) and the drain (D) open at the same time, Kenneth Boulding has nicknamed this equation the "Bathtub Law." The way the level of water in the tub changes depends on the rate of inflow (B) minus the rate of outflow (D), and it is through these inflows and outflows that the various aggregates of the system are connected.

The B term in one equation must consist of contributions from the D terms of the other equations, and the D terms of each equation must all be distributed among the B terms of the other equations. In other words, all the bathtubs must be connected and there must be no hidden leaks.

Consider, for instance, the arrangement of three tubs shown in Figure 5.1. Since there are three bathtubs, we start with three bathtub equations. The equation for tub 1 must be

$$N_1' = S - (P_1 + Q_1)$$

for S is the rate at which water is flowing into the tub (the birth rate, B) and $(P_1 + Q_1)$ is the water flowing out (the death rate, D)—through the drain (P_1) or by overflowing (Q_1). No mystery here.

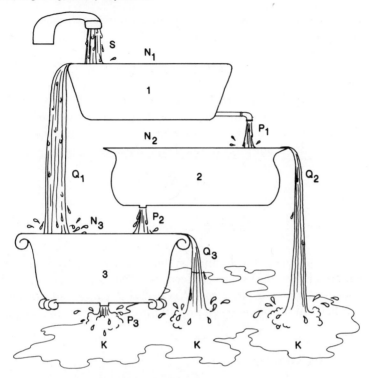

Figure 5.1. An illustration of the Bathtub Law, using real bathtubs.

S = Rate of flow from spigot
N_i = Amount of water in the ith tub
P_i = Rate of flow from ith drain
Q_i = Rate of overflow from ith tub
K = Rate of flow out of entire system

The other two equations may also be written down directly from the picture. For tub 2,

$$N_2' = P_1 - (P_2 + Q_2)$$

for the input to tub 2 is the drain of tub 1 (tub 1 is the "parent" of tub 2), and the output of tub 2 is the sum of what drains out and what overflows. Nothing sneaky here, either. For tub 3, similarly, the equation is

$$N_3' = (Q_1 + P_2) - (P_3 + Q_3)$$

Another way of looking at this system is to lump all three tubs together, so that the single aggregate variable, N, is the total of water in the three tubs. The equation for this system—for since there is only one aggregate, there is only one equation—is seen to be

$$N' = S - (Q_2 + P_3 + Q_3)$$

for the only source of water is the spigot and the only ways water leaves the system are by overflowing tubs 2 or 3 or by draining out of tub 3. Once we have written down this equation, we can use our knowledge of the lumping to check our three original equations, for

$$N = N_1 + N_2 + N_3$$

and

$$N' = N_1' + N_2' + N_3'$$

That is, any change in the lumped system must result from changes in one or more of the subsystems. Indeed, we find that when we add the three equations together, the terms P_1, Q_1, and P_2 cancel out, for they are "internal" flows in the lumped system and thus do not show in its equation.

To Solve or Not To Solve?

Briefly stated, the sentiment of mathematical elegance is nothing but the satisfaction due to some conformity between the solution we wish to discover and the necessities of our mind, and it is on account of this very conformity that the solution can be an instrument for us. This aesthetic satisfaction is consequently connected with the economy of thought . . .

It is for the same reason that, when a somewhat lengthy calculation has conducted us to some simple and striking result, we are not satisfied until we have shown that we might have foreseen, if not the whole result, at least its most characteristic features. (Poincaré 1952:31)

If, at the end of a lengthy calculation, we look back and ask how we could have foreseen the result, why not serve the economy of thought by looking back before we *start* calculating. We've now reduced the description of the bathtub system to a set of three differential equations, by three rudimentary applications of the Bathtub Law. We have even been able to check our work by writing an overall equation lumping the three subsystems and comparing it with the sum of the other three. This work required no great mathematical or physical knowledge. But before we can solve these equations, we will have to introduce the particular "laws of nature" into the equations, in place of the very general Ps and Qs. For instance, the rate of draining, P_1, from tub 1 might be a function of the diameter of the drain, the material lining its walls, and the pressure of the water. The water pressure, in turn, might be a function of the depth of the water. After we have built up a function describing P_1 in terms of all these factors, we can substitute its more complex but explicit form for P_1 wherever it appeared. In a similar manner, we can develop a formula for the rate of overflow, Q_1. Q_1 might be zero if the tub were not full— or we might wish to take splashing into account. Eventually, however, we can reduce each term of the equation to combinations of measurable attributes of the system in the familiar reductionist manner. Only then will it be possible to solve the equations.

Getting the details necessary to solve the equations may prove more difficult than drawing certain conclusions without solving the equations. For example, the exact equations governing overflow may be extremely complex, yet we may be able to avoid working them out by studying the system in cases where there is no overflow at all, that is, where $Q_1 = Q_2 = Q_3 = 0$. In that case, the overall equation

$$N' = S - (Q_2 + P_3 + Q_3)$$

becomes simply

$$N' = S - P_3$$

Furthermore, since the tubs have limited capacities, N cannot keep increasing forever, so at some point N' must equal zero. When that happens, then

$$N' = 0 = S - P_3$$

or

$$S = P_3$$

In other words, the rate of draining through P_3 must eventually match the rate of input under these assumptions, and if the input is steady, P_3 will eventually be steady too.

Many such conclusions can be drawn merely by inspecting the equations. Though training in the solution of differential equations can be helpful, the general systems thinker need not be intimidated. All too often, knowledge of mathematical techniques only impedes understanding. Techniques for solving differential equations are actually rather sparse. Great classes of potentially interesting equations cannot be solved analytically. Therefore, investigators who know too much mathematics may unconsciously trim their models to fit the equations they can solve.

Even solving a simple bathtub system like that in Figure 5.1 would require quite extraordinary mathematical methods. The P terms would probably contain factors such as $N_1^{1.85}$ making the equations "nonlinear." The overflow terms, the Qs, could not be expressed as continuous mathematical functions of the unknowns, for there is a sharp break, a "discontinuity," in the rate of overflow, which jumps away from zero when the tub fills to the brim. Thus, this trivial system of three bathtubs results in a system of nonlinear, discontinuous, differential equations. For such equations, there exists no general analytical method of obtaining even an approximate solution. Consequently, thinkers more at home with mathematics might never have formulated the system this way in the first place—in order to avoid such intractable equations.

Within a specialized discipline, the habit of constructing easily solvable mathematical models becomes deeply ingrained. Since the simplifying steps are seldom made explicit, students get the impression that nature has very obligingly arranged things in the form of continuous ordinary differential equations—just what they know how to solve. How convenient!

By exposing all the steps by which equations are built, the general systems thinker hopes to teach not only how to build better equations, but also when not to bother building them at all, let alone bother solving them. When you are a civil engineer, you study hydraulic network equations; when you are an electrical engineer, you study Kirchoff's laws; when you are an economist, you study input-output models; but whatever you are, you are studying bathtubs. Why do we keep hiding this fact one generation after another? Why don't we come clean?

Renaming Systems

There is another sort of procedure, applied to the counting of sheep or it is said formerly to the counting of an army. The sheep are originally all in one pen which is connected to another pen by a passageway so narrow as to permit the passage of only one sheep at a time. The sheep are all driven from the one pen through the passageway into the other pen, and counted one by one as they file through the passageway. So long as the walls of the pens remain impervious and lambs are not born in the second pen and we can be sure that there are no wolves in sheep's clothing this is a satisfactory procedure. That is, the procedure is satisfactory in general if we can be sure that what we are counting has, like sheep, certain of the properties of what we call objects. I suspect it would be extraordinarily difficult to make a list of necessary and sufficient properties to ensure this. I suspect we would find it difficult to avoid a circular proceeding and include as one of the properties that, when subjected to a counting procedure as above, we obtain consistent results. (Bridgman 1959)

One of the reasons we fail to see the similarity between bathtub systems drawn from different disciplines is that the equations we write and the pictures we draw are cluttered with mental junk which detracts from our understanding. We smile at the Victorian niceties of the bathtubs in Figure 5.1 for we know that griffin's feet have nothing to do with the trickling of water through this system. But the exercise of redrawing the tubs into one of our structure diagrams forces us to pay attention to the pith of the system—to assert what is, for us, essential.

Figure 5.2, for example, shows a redrawing of Figure 5.1 with certain details omitted. In the place of porcelain sarcophagi with griffin's feet we have three neat, crisp, and uninteresting rectangular boxes. In place of founts of water playing and splashing from tub to tub, we have neat, crisp, and uninteresting arrows. The rusty but trusty old faucet is gone, replaced by an antiseptic oval marked S; and even the puddles on the tile floor have been sponged up into another oval called K.

We have not gone to all this trouble just to make the tubs less interesting, but the loss of interest *is* a necessary intermediate step. At the level of precision of the Bathtub Law, all these gewgaws will only prove unusable and distracting detail. If we want to look at real bathtubs, we can go to a bathroom. For present purposes, we want to eliminate all secondary matter.

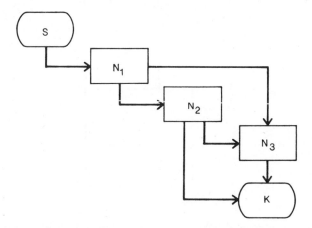

Figure 5.2. The three bathtub system of Figure 5.1 abstracted to a diagram of immediate effects.

As we know, even the oval shape of the S and K boxes is not *required,* but we wanted to call attention to the fact that they—the "source" of input S and the "sink" into which the output water drains K—are *outside* the system of interest. Consequently, they need be studied only insofar as they help to understand what takes place inside.

Another pictorial way we know of distinguishing "system" from "environment" is to enclose the system boxes within a larger box, as shown in Figure 5.3. Since the larger box engulfs the other three, its contents are what we called N in the previous section, where

$$N = N_1 + N_2 + N_3$$

The dotted box represents this relationship, and also shows graphically why the internal flows cancelled out in summing the equations. Indeed, we could erase the internal boxes and lines altogether and get the diagram shown in Figure 5.4—the same system at a different level of resolution.

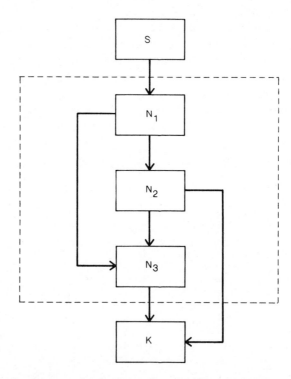

Figure 5.3. The diagram of immediate effects of Figure 5.2 with a dotted line enclosing the "system" and distinguishing it from the source and sink in the environment.

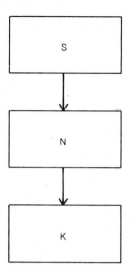

Figure 5.4. The diagram of immediate effects of Figure 5.3, with the subsystems no longer differentiated, but lumped into a single "system."

The central idea in all of these renderings of the bathtub system is that no matter how we draw it, it's always the same water. In other words, by the Principle of Indifference, we can relabel the water any way we choose, as long as it helps us fight the Principle of Difference. Each presentation is simply a different way of *partitioning* the aggregate of water molecules, so that *any* system that consists of "the same" individuals moving from one category to another can be represented by such a graph. For instance, the individuals may be identified by the position they occupy in space—the closest analogy to our bounded boxes. Examples of this type might be the circulation of electricity in networks, fluids in hydraulic networks, money in an economic system, deer in a forest, books in a library.

But the renaming of parts of the aggregate need not be on the basis of physical position, even though our structure diagrams are analogies with physical position. The library books may be classified according to who has checked them out—two books checked to Professor X may be in different places, and one of them may even be on Professor Y's shelf. All the people with incomes of $10,000 to $11,000 per year need not be in the same place to be named as a group and appear on a structure diagram as a rectangular box. In that case, the arrow does not mean "physically moves to" but "changes income to."

Or consider Figure 5.2 as the model of a junior high school. The S represents the students entering school, all of whom start in seventh grade, N_1. From seventh grade, most go to eighth grade, N_2, but some are skipped to ninth, N_3. Nobody graduates directly from seventh grade, but some do go directly to K from the eighth. The rest go to ninth grade and then graduate. Ultimately, everybody graduates from this school. Some may be kept in a grade for a long time, but nobody is ever put back to a lower grade, and nobody is ever kicked out. There is certainly no difference in *location* of the different grades—students of various grades may share classes, study halls, and recess periods. Yet we may use our physical analogy with undiminished effectiveness to get an overview of the movement of students through a school or an entire school system.

Whenever "what we are counting has, like sheep, certain of the properties of what we call objects," we have a special kind of system about which we can make some very general and powerful statements. We will call such a system a *renaming system,* because it is a system in which "objects" are not moved around, but merely given new names at various times.

The American kinship system is another example of a renaming system. People called "relatives" are moved into, out of, and around the system on the basis of such diverse criteria as "blood," "closeness," "sociability," and "accident." Schneider (1968) demonstrated this phenomenon in excerpts from his interviews with informants (A is the anthropologist; I, the informant):

A. Do you have to be close to someone to have them related to you?
I. Yes. You use the relationship. When it drifts away you are no more related. You see I went to one of my husband's cousin's bridal showers. It was for a first cousin's bride-to-be. You only meet all these people there. You meet them like at weddings or showers, or bar mitzvahs or funerals. For these things they call on you and I answer the roll call. (Shrugs her shoulders as if to say, "What could be more simple?") You walk in and you meet them all and half of them are pregnant, so you say, "How nice that you are going to have a baby, congratulations on becoming a new mother," and they say, "But I got two home already." So you see how it is.
A. So are these people related to you?
I. They are when you meet them like that, but when you leave them, they're not any more.
A. They are not related between weddings and funerals, but they are during them?
I. Yeah.
A. Have they ever been related to you except at things like weddings and funerals and bar mitzvahs?
I. Oh, sure, but they aren't now. You see this business of being related to someone has to do with sociability. There are social cousins.
A. Can you give me any kind of rule for the person who is related to you?
I. Well, they got to be sociable with you or they're not related.

A. All right, but some of the people you named are related to you by blood, right?

I. Yeah, you get them by accident. You can't do anything about them—and grandchildren are the bloodiest! (Schneider 1968:64)

As Bridgman says, "it would be extraordinarily difficult to make a list of necessary and sufficient properties to ensure" that our aggregate system can be considered a renaming system. Yet we do know some of the difficulties. One member mustn't suddenly change into two, as when lambs or grandchildren are born. No member must be mysteriously transformed into "nothing," as when a wolf drops its sheep's clothing disguise, or retains its disguise and eats a few sheep.

Generally, then, with a little checking we can satisfy ourselves that the aggregate meets these conditions and can thus be considered a renaming system. Or else we can see some fatal flaw, in which case the aggregate must be considered something else, and we lose some handy properties to work with.

The Renaming-Conservation Law

> Young children learn to speak . . .
> by constantly using words to bring things *into their minds,*
> not *into their hands.* (Langer 1942:109)

What does it buy us to have a renaming system to work with? Most important, we think, is that a renaming system assures us of the ability to count with consistent results. If the system indeed is equivalent to one in which members neither reproduce nor disappear, there will always be the same total count of members, no matter what renamings take place. In other words, whatever changes the state vector

$$(N_1, N_2, \ldots N_g)$$

might undergo, the *sum* of all the N_i *remains constant.*

This property may be stated in the form of the *Renaming-Conservation Law:*

Every renaming system has an overall conservation law.

This law, in reality, is simply a corollary of the Principle of Indifference. No matter how we rename the objects, we don't expect to change their number. "Sticks and stones may break my bones, but names can never make me disappear or multiply."

Sometimes we go to great lengths to ensure that our system has the renaming property. In a library, the circulation of books doesn't change their number, though in a real-world library, some of the circulating books get lost. To preserve the renaming property for a study of library circulation, we might have to add a category of "lost books." If this category eventually gets too big, however, we may be most unsatisfied with our model of the library—or with the library itself.

The Renaming-Conservation Law permits us to write such simple bathtub equations, at least in gross structure, and to check our work by adding all the equations together. Both Boulding's bathtub and Bridgman's sheep are physical analogies against which we can test an aggregate system for the renaming property. In the tubs, the individual atoms of water are very small but very numerous. We don't imagine counting them one by one, but as "flowing" continuously from one tub to the other. The tubs, then, are the archetypical *continuous* system, which we would usually model by *differential* equations. In these equations, we *pretend* that the water is infinitely divisible, though we know at the same time that it is composed of discrete molecules which must be conserved.

The sheep in pens are less numerous, but bigger than atoms. They are big enough so that people can deal with them as individuals. Indeed, a good shepherd has a name for every one of the hundreds of sheep in the flock and can determine not only that one is missing, but which one is missing. The sheep are the archetypical *discrete* system, which we would usually model by *difference* equations, such as

$$N_{t+1} = N_t + B_t - D_t$$

instead of the differential form

$$N' = B - D$$

Ultimately, the size, N, of an aggregate is a *count* of members, but if the members are very numerous, we may replace this count with an *approximate* count, which we call "measuring." When we can actually count, we use the *difference* equation, because that's based on counting directly. A *differential* equation, on the other hand, takes us to a higher level of abstraction, since we're now measuring change in terms of tiny differences in which the identity of individual members has disappeared. You can count sheep, but only a desperate insomniac would count water molecules.

Whether we compute exactly, by counting, with the difference equation, or approximately, by measuring, with the differential equation, we have a conservation law. We can add all the separate equations for the individual aggregates and get the overall equation for the undifferentiated total aggregate. (Don't let the two words "differentiated" and "differential" confuse you by their similarity. Indeed, the "differential equation" is used when the individual members of an aggregate are *not* "differentiated," but, rather, are part of one continuous mass.)

Having a conservation law helps us in other ways. Sometimes we are working in the opposite direction. We know that there is a conservation property and are looking for an explanation. Not all conservation laws come from renaming systems, but many of them do, so it's usually a good idea to look for some renaming hidden behind apparent conservation.

Consider the case of Roussillon, the French commune whose population stayed at 700, even though only 275 of the original inhabitants remained. In searching for an explanation of this constancy (and it *is* constancy that has to be explained), what renaming system can we discover? Well, it could be that some of the 425 *are* the same people with new names, such as newly married women or former collaborators. Indeed, it's quite likely that some of the people we thought were different are actually the same, and that's a worthwhile demographic insight in itself. Turning it around, it could also be that some of the people with the *same* names are different *people*. This, too, is a discovery many peasantry demographers have made, late in their work, for many villages have dozens of duplicated and reduplicated names.

But these two factors are unlikely to account for all 425 changes, or all 275 nonchanges. Where else can we look? We could try to construct a somewhat more *encompassing* system that takes in the 425 who left and the 425 who arrived, the way we did when we added source and sink to the bathtub or school systems. Let's consider some facts about the *region* in which Roussillon is found.

About half the immigrants to Roussillon come from neighboring communes, and more than one quarter of the emigrants move to these same communes. Roussillon, like some other peasant communities, shows large turnover in population, essentially an exchange of people among several villages or communes in the region. By taking the large unit, the region, into the system, we create a conservative system from what might have seemed incredible volatility if only one community were studied.

Sometimes the insights come from finding a *different* renaming system that has some special relationship to the system we are studying. Although there were many changes in *people* in Roussillon from 1946 to 1954, the physical structure of the village may not have changed much.

It was just after the war. Living space was scarce. Therefore, we might speculate that, though the *people* were changing rapidly, the *dwelling units* might have been pretty much constant.

When we investigated this possibility, we first learned that we were grossly mistaken, because in 1951, one-third of the houses in Roussillon were vacant (Wylie 1964). But reading further, we discovered that in spite of these vacancies, there was a housing shortage, and young villagers had to postpone marriage or move in with parents. The houses were vacant because their owners were unwilling to accept the low rent ceiling fixed by the government. Because it was hard to evict a tenant, landlords preferred to wait until they had an opportunity to sell the house.

By 1959, a new market of house buyers appeared—not villagers, but urbanites from Paris and elsewhere. Taking advantage of the depressed housing market of dying rural communities, they bought property cheaply for vacation use. As a result, real estate prices skyrocketed. As Roussillon "developed its new function as a resort town the last vestiges of the former agricultural center . . . almost disappeared" (Wylie 1964:342).

Let's follow the course of reasoning as it might have happened to someone researching a village like Roussillon. The observed conservation of people leads us to think about something *else* that might be conserved that would indirectly lead to the conservation of people. We think of houses. This gives a direction to our research. Variable numbers of people could live in the same unit, leading us to examine living patterns. We might discover that people in this culture have relatively rigid ideas about how many people should occupy a given space, or we might find that although the number is variable, the *average* is pretty constant.

Armed with this information, we are ready to think about the system in a new way. Instead of renaming people, we rename living spaces. And what is the "name" of a living space? Quite simply, the name of the *person*, or family, living in it. Thus, as people move in and out, they are *renaming the living spaces*, from our point of view. No wonder, then, that the total number of people was relatively constant.

When we try to validate our living-space model, we may run into the problem of vacancies, which is a challenge to the model. But the model gives us a set of questions to ask, and leads us to discover the existence of large amounts of unused space, which might otherwise have escaped our notice, particularly in historical research. Knowing about these spaces, we are prepared for the "sudden" change in population that occurs when a slight shift in property values takes place, or when the rental restrictions are relaxed.

As we pursue these questions, we'll be helped by the simplicity of our hypothesis. There can be other models which would give a stable population without being renaming systems. Indeed, their possible number is unlimited, which is one of the principal reasons we want to start with something simple. Although every renaming system has a conservation law, not every conservation law comes from a renaming system.

Later in this book we shall examine many other systems that conserve some variable. In general, they are more complex than renaming systems. For that reason, we don't want to spend a lot of time working out a grand scheme of interlocking feedback mechanisms only to discover that the phenomenon we are trying to explain is a straightforward consequence of a system with the renaming property. We'll try the simplest first, and seek the complicated only if forced into it.

Simulation and Conservation

It is our case that for many phenomena that affect an individual in his everyday affairs—and for many others that affect him only by arousing his curiosity—that individual needs a mental model of the phenomena. Many such models are formed quite unconsciously for a daily practical purpose. Such models need not be correct; in fact, one does not so much ask whether or not the model is correct as whether it adequately or inadequately fulfills its purpose. When we travel to work, we unconsciously use a flat-Earth model of the terrain, as we think most people do on their ordinary journeys. In estimating when we are to arrive, we do not take into account the increased distance we shall have to travel on account of the curvature of the earth; in our imaginations we could see our office from home if it were not for the intervening hills, trees, and haze. . . . we simply say that all models are "right" when applied to a purpose for which they are useful. (Coats and Parkin 1977:5–6)

We have the bathtub and the sheep analogies to help us think about renaming systems, but because the conservation property is so helpful, we could use a few more analogies. To the younger generation of scientists, who may never have seen a flock of sheep and who are unlikely ever to have used a bathtub, a computer simulation might be more familiar.

Probably the most direct way to simulate an aggregate system on a computer is to make each computer memory cell represent one member of the aggregate. Then we can do symbolically to the memory cell what we believe the system does to the individual, with much greater convenience and much less expense. For instance, if we are simulating the library circulation, each memory cell could represent one book. Handling a collection of 1,000,000 memory cells ought to prove less laborious, cheaper, faster, more accurate, and less dusty than handling 1,000,000 books.

In each cell, we could store any information we'd like to follow concerning the circulation of books. Most important, though, would be a code to indicate the "state" of the book at any instant of time, such as (Figure 5.5):

> SH = On the shelf
> BI = At the bindery
> RE = On reserve
> FA = Checked out to faculty
> ST = Checked out to student
> HO = On hold, waiting to be checked out
> VA = In the vault for rare books
> UN = Whereabouts unknown

The computer program simulating the circulation of books would change the code in a memory cell when the status of the simulated book changed. A return of a book taken out by a faculty member would result in a change from FA to SH, unless there is a "hold" waiting for the book. In that case, its status is changed to HO when it is returned. It stays HO until a simulated student checks it out. Then it changes to ST. In this way, the program may pause at any time and count the number of books

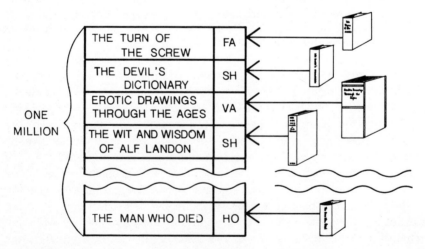

Figure 5.5. In a primary simulation, each "cell" of computer memory would hold the information concerning one member of the aggregate, in this case, one book.

(memory cells) in each status (with each code). If the program has accurately simulated the behavior of librarians, faculty, and students, these totals will accurately simulate the state vector of the real library.

A simulation of this type, in which each memory cell represents one member of the simulated aggregate, can be called a *primary* or *one-to-one simulation*. If, further, the number of cells is fixed throughout the simulation, we can call the simulation a *fixed primary simulation*. In our library example, the simulation can be fixed only if we do not consider acquisitions during the period simulated and if we account for losses by providing a catchall category like UN for unknown.

A fixed primary simulation is probably the easiest kind for which to write a computer program. The programmer just lays out the cells, allowing one for each member of the aggregate and sufficient space within each one for any information needed about that member. Then all the rules for *moving a member from one state to another* are written down in the appropriate programming language and the job is done (with a little program testing, of course).

For those who have had a little experience programming computers, it's easier to imagine a fixed primary simulation than it is to count sheep. If we *can* imagine such a simulation, *then the system simulated must be a renaming system*. Why? Because that's exactly all a fixed primary simulation does—it renames things.

Imagining a fixed primary simulation is another kind of analogical thinking, just like imagining sheep or bathtubs. Just as we don't actually have to handle sheep or take baths to use these analogies, neither do we have actually to *program* the simulation. For instance, in the case of simulating a bathtub-water system, we could *imagine* each memory cell containing the position of each single water molecule. We would *not* actually need a computer with 10^{25} memory cells.

Even in the case of the library system, an actual simulation of 1,000,000 books might be far too expensive on available computers. All we need to know is that the library circulation system is isomorphic with a computer memory of 1,000,000 cells and a program for transforming those cells in the same pattern that books are moved or reclassified in the library. Then we know there is a law of conservation of books.

The computer system is a *model* of the library system. Our mental image of the computer system, in turn, is a *model of the model*, and therefore a model of the library system, as well. This mental model can actually be extremely vague and imprecise—like the image of the flat earth—and yet still be useful for thinking about the library system. If computer modeling is useful because it can save us the trouble of working with the real system, just think how much more trouble we save by working with a model of the model, and not programming the computer at all.

In real simulations of large aggregates, we seldom use a primary simulation approach. Instead of having 1,000,000 cells for 1,000,000 books, we use one cell to keep a count of each *category* of book. That is, we keep one cell for each element of the state vector. With eight circulation categories, we would need eight cells, each containing a count (Figure 5.6).

To simulate a system of three tubs plus one source and one sink, we'd need only five cells, even though there might be 10^{25} water molecules. The saving in computer resources is enormous, but there is some loss of precision. Whereas the primary simulation concentrates on the water molecules—the *individuals*—the secondary simulation concentrates on the tubs, the *aggregates* in the state vector. Thus the terms primary and secondary.

Notice, however, that what are aggregates in one system can become individuals in another as we change our level of resolution. In a simula-

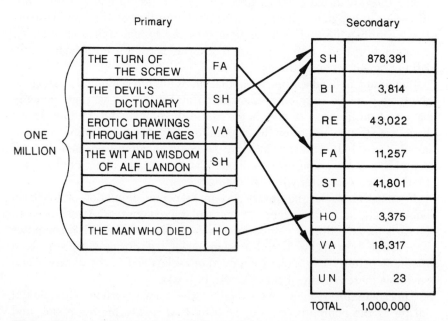

Figure 5.6. A primary and a secondary simulation of the library system. In the primary simulation, there must be 1,000,000 cells for 1,000,000 books. In the secondary simulation, each cell represents one *category* of books, along with the number of books presently in the category. No matter how many books there are, there will be only 8 cells in this simulation, because there are only 8 categories. If there are 1,000,000 books, the numbers *in* these 8 categories will *sum* to 1,000,000.

tion of the water system of a neighborhood, the system of tubs in one house merely becomes part of the water usage of one house out of a large aggregate of houses. The level of resolution that was a secondary simulation with respect to water molecules becomes primary with respect to the neighborhood. Thus the terms are only useful in a relative sense, to compare one simulation model with another (Figure 5.7).

In the primary simulation, the one-to-one correspondence between a fixed number of cells and the individuals of the aggregate *guarantees* the conservation property. In a secondary simulation, conservation is *not* guaranteed in this way, since the one-to-one correspondence between parts of the simulation and parts of the system is destroyed. The number of *tubs* is conserved, but for the moment we're not interested in tubs. The

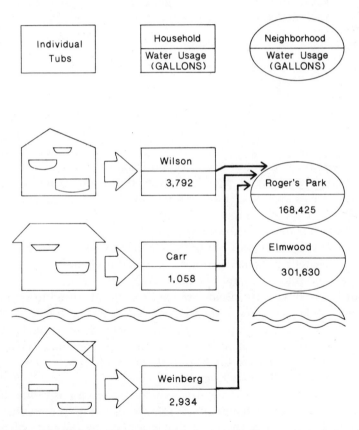

Figure 5.7. The bathtubs lumped together by household are a secondary simulation with respect to the initial bathtubs, but a primary simulation with respect to the larger neighborhood simulation.

overall aggregate *might* be conserved, but more work is needed to *guarantee* conservation.

A fixed primary simulation can simulate only renaming systems, but a secondary simulation could be used to model any kind of discriminated aggregate in which the number of tubs—the number of elements in the state vector—is fixed. For example, an exponentially growing aggregate would need an exponentially growing number of memory cells for a primary simulation. For a secondary simulation, the number of cells would not depend on the number of individuals, but only on the number of *types* of individuals discriminated.

When even the number of *categories* is not fixed, sometimes a tertiary simulation can be used, where each cell keeps count of the number of categories in a certain grouping of categories. Each level of simulation represents a level of resolution of thought about the aggregates, whether the simulation is actually performed or not.

Tracers

Skill acts, as it were, a radioactive tracer in the blood stream of migration, and . . . we would do well to develop the study of it as an instrumental technique. (Thistlethwaite 1964)

Even in systems, obeying no conservation laws, there may be subaggregates of subsidiary interest that *are* conserved. Reasoning about nonconservative systems may be facilitated if we can shift our point of view and focus for a time on subaggregates that *are* conserved and that are related in some known manner to subaggregates that are not.

The conserved aggregates can be called "tracers"—by analogy with the technique whereby radioactive atoms are bound up in molecules ingested or injected and traced through the body. Although the original molecules may not be conserved, the atoms are, and can thus be used to probe the body's metabolic processes. Similarly, migrants may be traced by traits that are conserved through a variety of migrations. Small variants of dialect may pin down an origin to a region, district, commune, village, or even to a particular block in the heart of London, as Professor Henry Higgins could do with Eliza Doolittle.

By similar techniques, we may trace the origin of an upswelling of sea water to its source by its temperature, salinity, or isotopic composition, or we may deduce the populations of fish in a pond by the rates of disappearance of foods. We may reason forward from known sources to unknown sinks, or backward from known sinks to unknown sources.

The more different conserved aggregates there are in our system, the more likely we are to be able to make deductions, since each conservation law allows us to write down an extra relationship among the elements of the system. For this reason, one of the first generalist questions is always, "What is conserved?"

Yet this rage for conservation often merely exacerbates a difficult situation. Just because we *see* something as an aggregate, or call it an aggregate, doesn't mean it will have conservation properties. For that, it must either be a renaming system or have a very special kind of program, neither of which need be true for an arbitrary aggregate chosen by us.

A sterling example of such an unrewarding quest has been the search for something called "information" which would be conserved along with the two biggies of physics, matter and energy. For many years, researchers have sought fame and fortune in the form of some sort of conservation law for "information." Anatol Rapoport (1968) noted that a 1953 bibliography of "information theory" contained some 800 entries. Twenty-five years later, a similar bibliography would certainly contain ten times that number—all in the heretofore vain hope of finding a "thing" called "information" that would be conserved.

About this quest, Rapoport observes: "It is misleading in a crucial sense to view 'information' as something that can be poured into an empty vessel, like a fluid or even like energy." Nor, we might add, can it be counted like sheep, for indeed, our intuitive view of "information" typifies precisely those "things" for which general conservation laws *cannot* be found—unless there is a very special program at work to make conservation happen. Information can be duplicated readily, with minuscule cost in energy, even faster and easier than sheep can breed. Information can disappear, such as when the last existing copy of an ancient book crumbles to yellow dust on a forgotten library shelf.

Yet there are many *aspects* of certain kinds of information that are conserved, and these aspects can provide useful tools for studying systems of information. For instance, typographical errors in printed information are copied when the information is copied, at least by mechanical means. Because these errors are conserved, they can be used to trace the origins of derivative documents, just as skills, or accents, can be used to trace the migration of people. Indeed, publishers of tables and other reference works often make *intentional* errors just to be able to *prove* another work was copied from theirs. A cartographer, for instance, might invent a town called Dacron, Ohio, and put it on his map. He can be quite certain that any other map having Dacron in the same place was copied, possibly indirectly, from his.

It would seem better, then, to spend our time carefully designing conserved aggregates than in trying to discover the one sensational one that will win the Nobel Prize. At least this is safer than making them up, casually, as we go along, such as in this typical quotation:

There is also a possible connection between the level of mortality and the amount of emotional energy that parents invest in each of their children . . . A reduction of mortality encourages parents to place more libido in the existing children and thus should reduce their desire to have an additional child, since they have limited amounts of emotional energy. (Heer 1968)

The "possible connection" here is argued on the basis of something called "emotional energy." By the association of names, we are supposed to infer some sort of "Law of Conservation of Emotional Energy." If we make this inference, the rest of the reasoning seems relatively sound, since it has the same pattern as all conservation reasoning. What goes one place surely cannot go somewhere else, can it?

It can, of course, if it is something like "information" or "emotional energy," which, for all we know, is *increased* by being "spent." Whether or not there even exists such an aggregate as "emotional energy," we don't know, nor do we know if this presumed aggregate obeys a conservation law. Although the metaphor is appealing, it's probably incorrect. If it's incorrect, it could still be good systems thinking—if it led to something else of interest. Yet as it's used here, it's more likely to *discourage* further explanation, and that would certainly classify it as bad systems thinking. Now *there's* a category that threatens to grow exponentially!

QUESTIONS FOR FURTHER RESEARCH

1. *Zoology*

If there is anything we take for granted more than rocks and trees, it is our own bodies. For example, the fluid medium we call blood must be maintained or we die. Discuss the circulation of the blood as a bathtub system, paying particular attention to the mechanisms ensuring that the total blood volume remains roughly constant—the so called "haemostatic" mechanisms. (Reference: MacFarlane 1970).

2. *Economics*

Gold is a product which is not easily created or destroyed. Moreover, it is highly valued in most cultures, so it is not very often lost. Consequently, the movement of gold in international trade should prove an ideal case of a bathtub system. Unfortunately, things are not quite so simple, as discussed by Morgenstern in *The Validity of International Gold Movement Statistics* (1955). Using this source and others, try to diagram the flow of gold in international trade as a bathtub system, and discuss the difficulties involved.

3. *Anthropology*

Evolutionary thinking has recently stimulated anthropologists to consider the "developmental cycle in domestic groups" (Goody 1971). The domestic group

must remain in operation over a stretch of time long enough to rear offspring to the stage of physical and social reproductivity if a society is to maintain itself. This is a cyclical process. The domestic group goes through a cycle of development analogous to the growth cycle of a living organism. The group as a unit retains the same form, but its members, and the activities which unite them, go through a regular sequence of changes during the cycle which culminates in the dissolution of the original unit and its replacement by one or more units of the same kind. (Goody 1971:2)

To what extent can the domestic group usefully be modeled as a renaming system?

4. *Anthropology*

Cultural anthropologists who study peasant societies have become convinced of the necessity to set their investigations in a larger context than simply the particular village under study. As Eric Wolf wrote in 1966, "What goes on in Gopalpur, India or Alcalá de la Sierra in Spain cannot be explained in terms of that village alone" (1966:1). The introduction to a recent collection of papers on small-scale societies in Europe elaborates:

Political, religious and economic relationships, say, in an Italian village, clearly do not exist in isolation at a local level. They are influenced by relationships and processes that lie beyond the community at regional, national, and even supranational levels. Terms such as group, village, community, culture, society have been used to indicate socially significant entities. Concepts like brokerage, encapsulation, penetration, folk-urban, great tradition and little tradition, absorption, and acculturation have been brought forward to deal with aspects of relations between these entities. These terms and concepts, which are used by most anthropologists as scientific instruments, were largely developed to describe and analyze relations in small-scale, fairly isolated communities. To polarize part and whole, micro and macro, community and nation in the study of complex European societies by reifying them as separate categories does violence to the nature of the dynamic relationships between them, and the meaning they have to the people involved. Yet here lies the rub. Anthropologists have done little to systematize their thinking on the nature of these relationships to avoid this static polarization and reification. This is the central problem to which this volume is addressed. (Boissevain 1975:9)

As a general systems thinker, how would you approach this "central problem"? What principles and tools would you suggest?

6

Programs for Models of Differentiated Aggregates

In the description of machinery by means of drawings, it is only possible to represent an engine in one particular state of the action. If indeed it is very simple in its operations a succession of drawings may be made of it in each state of its progress, which will represent its whole course; but this rarely happens, and is attended with the inconvenience and expense of numerous drawings. The difficulty of retaining in the mind all the contemporaneous and successive movements of a complicated machine, and the still greater difficulty of properly timing movements which had already been provided for, induced me to seek for some method by which I might at a glance of the eye select any particular part, and find at any given time its state of motion or rest, its relation to the motions of any other part of the machine, and if necessary trace back the sources of its movement through all its successive stages to the original moving power. (Babbage 1826:346)

In the great age of machinery, the nineteenth century, people first began to realize that it was possible to build things you couldn't really understand. In parallel with the invention of millions of new machines, but always lagging behind, was the invention of thousands of notations by which the inventors hoped to understand the workings of their machines. No history of these notational inventions has yet been written. Probably none ever will be written, for though we can usually reconstruct most of the functioning and purpose of an ancient machine, the symbolic systems which describe them remain undecipherable.

In our century, the great notational concept is also based on a machine—the mathematical machine that Babbage spent his life trying to build. The computer, and especially its programs, is in many ways the system that Babbage and thousands of others needed for the description of machinery. The computer program analogy is the great conceptual tool of our time. Whether or not programs will be decipherable in the twenty-first century, we can't say, but our job is to use them to make a variety of "machines" more decipherable today.

Varieties of Programs

To set up equations means to express in mathematical symbols a condition that is stated in words; it is translation from ordinary language into the language of mathematical formulas. The difficulties which we may have in setting up equations are difficulties of translation. (Polya 1945:160)

In the fundamental systems equation:

$$S_{t+1} = F(S_t, I_t)$$

F stands for the *program,* or transformation, by which the state at one instant of time, combined with the input at that instant, produces the following state, S_{t+1}. We've seen that when S is a *vector,* this equation can be thought of as a single vector equation or a collection of scalar equations. But "program" is a far more general concept than "equation" and in this chapter we want to examine some of the forms that programs take when we try to model various discriminated aggregate systems.

Why is "program" a more general concept than "equation"? In some abstract mathematical sense, the two are equivalent, but practically speaking, equations are limited to relatively well-behaved mathematical operations. What do we mean by "well-behaved"? An example of poorly behaved is the best way to illustrate this concept—anything that isn't poorly behaved is "well-behaved."

Let's go back to our polling example and imagine that we are simulating the polling of 1000 people in the village. We decided to do a primary simulation, so that one of the 1000 cells in our computer memory is labeled "Jurgen B. Donahue." Now Jurgen B. Donahue is a 93-year-old recluse who only goes out of his shack to shop for molasses, coffee, and brown rice. He's deathly afraid of strangers, so if he starts for the grocery store and spots the polltaker out front, he hides behind the mailbox until the way is clear.

Evidently, Jurgen B. Donahue's eccentric behavior is going to affect the chances that the polltaker will ever get in touch with *everybody* in the village. The polltaker *might* catch him by accident, or by some counterploy of his own. In any case, the polling of Jurgen—and several other eccentrics in the village—is not going to be easily modeled by a mathematical formula.

In a computer program, however, we can model this behavior to any level of detail we care to go. It won't always be easy, but at least it will be more straightforward than trying to concoct a formula for this poorly behaved individual. The program can contain, if necessary, a separate set of instructions—a "subprogram"—for each individual in the village. For Jurgen B. Donahue, part of the subprogram might read:

IF MOLASSES IS LOW OR COFFEE IS LOW OR BROWN RICE
IS LOW, START FOR THE STORE.

In the procedure for going to the store, the program might read:

IF ANY STRANGER IS IN FRONT OF THE STORE,
    HIDE BEHIND THE MAILBOX AND KEEP WATCHING.

More and more such detail can be placed in the computer program until we have an uncanny match that's almost more like Jurgen B. Donahue than Jurgen B. Donahue himself.

Would it be worth the trouble? Probably not. Computer programmers often express a feeling of superiority to mathematicians, because they can much more easily construct such detailed models. All the same, it would still be surpassingly difficult to model 1000 distinct individuals at this level of programming detail. Except in some rare instances, such as simulating the judges on the United States Supreme Court, it's unlikely to be worth the expense. It will probably be easier just to send the polltaker out with instructions to get every last person, no matter what. The money saved on programming can be offered as an incentive for getting the job done.

There's always a fundamental limitation to *any* simulation—the cost of actually working out the problem on the system itself, rather than on a simulation model. Oh, yes, there are some simulations like a doomsday bomb or the creation of the universe that you won't or can't carry out. But in practice, the simulation is only worth doing if it's *cheap*. And that's where the mathematicians can sometimes outshine the programmers.

If the system can be modeled with sufficient accuracy by a simple set of equations, then the cost of each behavioral simulation ought to be low. Either the equations can be solved analytically, or a general-purpose computer program can be run with little or no special programming. For instance, in "linear programming" models, the relationships among the "activities" (components of the state vector) are all expressed as linear algebraic equations or inequalities. The "program" here can be completely described by writing down the values of the coefficients in these equations (Danzig 1957). Thus by obtaining the values of approximately g^2 numbers (a $g \times g$ matrix), we can use a standard "linear programming package" to obtain all sorts of information about the state vector,

$$(A_1, A_2, \ldots, A_g)$$

where each A is an "activity," or what we call an "aggregate."

In "input-output models"—or "Leontief models"—the relations among various productive sectors of an economy are similarly represented in a tableau of g^2 numbers. Each number shows how much of the output of one productive part goes into the input of another (Leontief 1941). If we substitute "bathtub" for "productive sector," the analogy is exact, except the input-output model uses simpler (linear) equations than the hydraulic system could accept.

In electrical networks, each component "connects" with one or more other components, and the connections can be described by another g^2 tableau (Zadeh and Desoer 1963). Over and over, the relationship among the "parts," "activities," "sectors," or "subaggregates" of a differentiated aggregate system can be related by equations taking the same structural form, a tableau of g^2 elements.

In other cases, however, the relationships are not necessarily so quantitative and may be represented only by diagrams, such as we used for illustrating the Bathtub Law. But even when there are diagrams, we can sometimes use them as programs to carry the behavior of the system forward in time, at least in a qualitative sense.

Transitive Closure—The Diagram of Possible Effects

Visual thinking pervades all human activity, from the abstract and theoretical to the down-to-earth and everyday. An astronomer ponders a mysterious cosmic event; a football coach considers a new strategy; a motorist maneuvers his car along an unfamiliar freeway: all are thinking visually. You are in the midst of a dream; you are making order out of the disarray on your desk; *you* are thinking visually. (McKim 1972)

Reasoning with the aid of diagrams is perhaps one level above verbal reasoning. It shares many pitfalls with verbal reasoning but also provides a useful complementary tool. Our language system seems based on our hearing, which is essentially one-dimensional. Diagrams and pictures interact with our visual system, in which two-dimensions are available. Therefore, fallacies in one system are often exposed in the other.

Just how the verbal and visual systems work is only partially understood. Reasoning with diagrams is a poorly developed art. Linguists have given us many years of insight into verbal processes but still leave worlds to be done. Only recently has anybody studied visual reasoning, and then mostly as an abstract mathematical discipline—graph theory. And although graph theorists (Berge 1973) have studied the formal properties of structure diagrams, what we really need is some work by the psychologists.

One useful idea which comes from graph theory is that of the *transitive closure* of a graph. We obtain the transitive closure from the structure diagram by adding an arrow (i, k) whenever there are arrows (i, j) and (j, k) for some j (Figure 6.1). The process is continued until no new arrows can be drawn, at which point the diagram is *closed*. The closure of Figure 5.2, our three bathtubs, is shown in Figure 6.2.

The reasoning behind the transitive closure of the structure diagram of an aggregate system is that if some substance can go from i to j and from j to k, it can go from i to k. If we find a pollutant at one point in a water system, we can trace backward along the arrows that go to k to find all potential sources. The transitive closure is thus seen as the graphic equivalent of functional reduction. If k depends on j and j depends on i, then k *may* depend on i. All the dangers of functional reduction are inherited by transitive closure.

Ashby (1956) has called the transitive closure by the suggestive name, "diagram of ultimate effects," but in view of the potential difficulties, it would seem more prudent to use the name, "diagram of *possible* effects." In the case of aggregate diagrams, we might use the name "diagram of possible flows," or "diagram of possible renamings," since the diagram shows all the places from which an element in a given box might have come.

In a way, it is even more prudent to call the transitive closure the "diagram of *impossible* effects," calling attention to those places where lines are *missing*. Thus in Figure 6.2 it is impossible for K to affect anything, since no arrows go out. This may be taken as the definition of a "sink." We should be wary, however, for sinks occasionally "back up." In the same way, a "source"—a box such as S which no arrows can reach—sometimes "dries up" or is "contaminated" by some sort of backwash that we didn't reckon on.

This kind of reasoning back to sources and forward to sinks explains why we must be so circumspect about accepting nonconserved aggre-

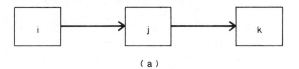

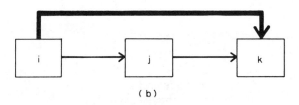

Figure 6.1. The transitive closure of a graph is formed by adding an arrow from i to k whenever there are arrows from i to j and from j to k. The heavy arrow makes b the transitive closure of a.

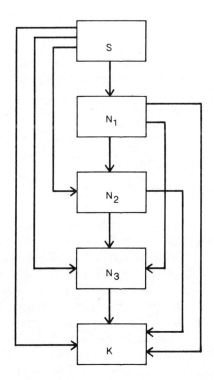

Figure 6.2. The transitive closure, or diagram of ultimate effects, of Figure 5.2—the three bathtub system.

gates as if they are conserved. A path along which a single idea can move is potentially as disruptive as a television cable. A single pregnant spider floating from one island to another on a coconut is potentially equivalent to an invasion. Thus, with nonconserved aggregates, it will be almost impossible to reason either forward or backward through the transitive closure and assert definitely that something will or will not—did or did not—affect something else.

Structure Matrices

> All the lonely people—
> Where do they all come from?
>
> Lennon and McCartney
> from "Eleanor Rigby"

Although graphs and their analogies with physical space are suggestive to look at, they are difficult to manipulate. When we want to make projections of possible effects, flows, or classifications forward in time, another method of representing connectedness may be used—the *structure matrix*. "Matrix" is simply the mathematical name for a two-dimensional tableau or array of symbols. For example, a mileage chart may be arranged in matrix form such as

	Chicago	New York	Seattle
Chicago	0	840	2064
New York	840	0	2904
Seattle	2064	2904	0

This is a "three-by-three" matrix, with the headings given only for convenience—only the numbers are considered part of the matrix.

If the entire matrix is called M, the individual members may be named by giving a pair of subscripts to M. In this table, M_{cn}, read as "M-sub-c-n," could stand for the distance *to* Chicago (c) *from* New York (n). If we number the rows and columns each from one to three, the same element could be designated as M_{12}, read as "M-sub-one-two." Some general element could be given variable subscripts, as in M_{ij}, read as "M-sub-i-j," and pointing to the "ith row" and "jth column." Such variable subscripts are useful in making certain statements about the matrix, such as, in the above matrix, that $M_{ij} = M_{ji}$ for all values of i and j.

When $M_{ij} = M_{ji}$ for all i and j, we say that the matrix is *symmetric*. In the case of a mileage table, we expect that the matrix will be symmetric, for the distance from Chicago to New York should be the same as the distance from New York to Chicago—though it might not if there were one-way roads. Certainly not all matrices need be symmetric, even for traveling. If the entries in the matrix are travel *times* rather than travel *distances*, it might well be nonsymmetric. To go from Denver to Omaha may take less time than to go from Omaha to Denver, since one way is mostly up hill and the other mostly down.

Almost any matrix tells us something about the structure of some system it represents. Our road mileage matrix tells us that all three cities are connected by roads, which might not be true if we had Sidney and Tokyo in the same chart. In some cases, we are interested in using the matrix form only to display the grossest structural facts about a system, such as whether it is *possible* to get by road from one city to another. In that case, we only need two symbols in the matrix, for which we conventionally use "one" and "zero" in the following way:

0 = no connection exists between these two
1 = connection exists between these two

Thus, we might have the following structure matrix indicating road connections among four cities:

	Tokyo	Osaka	Sydney	Perth
Tokyo	1	1	0	0
Osaka	1	1	0	0
Sydney	0	0	1	1
Perth	0	0	1	1

The same method may be used in transcribing structure diagrams into structure matrices, because the diagrams, after all, are analogies to connection in physical space. The interpretation we place on zero and one may be

0 = no flow can go from one to the other
1 = flow can go from one to the other

or possibly

0 = no renaming can be made from one to the other
1 = renaming can be made from one to the other

or, finally,

0 = one cannot affect the other
1 = one can affect the other

Thus, for example, if we transcribed Figure 5.2 (the bathtub diagram) we would have the matrix:

	S	1	2	3	K
S	0	0	0	0	0
1	1	0	0	0	0
2	0	1	0	0	0
3	0	1	1	0	0
K	0	1	1	1	0

which we would write without the headings as

$$
\begin{matrix}
0 & 0 & 0 & 0 & 0 \\
1 & 0 & 0 & 0 & 0 \\
0 & 1 & 0 & 0 & 0 \\
0 & 1 & 1 & 0 & 0 \\
0 & 1 & 1 & 1 & 0
\end{matrix}
$$

This form contains all the *essential* information from the structure diagram without any trace of extraneous material. Not only have we eliminated griffin's feet, but also shapes of boxes, layout on the page, lengths of lines, and anything else that might assist us in interpreting the model. But as before, stripping away this suggestive information may make other information move vivid. We see, for instance, that *row* S contains entirely zeros, which means that S is indeed a "source," since there is no arrow going into it. Similarly, a "sink" can be recognized by a *column* with all zeros, as we indeed see for K.

Time Scale and Continuity

> Out upon it, I have lov'd
>   Three whole days together;
> And am like to love three more,
>   If it prove fair weather.
>
> Time shall moult away his wings
>   Ere he shall discover
> In the whole wide world agen
>   Such a constant Lover.
>
>   Sir John Suckling
>   *Song*

How much love is a lot? How long must a constant be constant? When studying the *general* question of the effect of one aggregate on another, we take no account of time. The ones and zeros only tell us that an effect is possible or impossible without giving a clue about how long it may take for an effect to be felt.

A diagram of immediate effects is merely a sketch of the general structure connecting the parts of a differentiated aggregate system. A matrix can translate the information on that diagram into numbers that can be manipulated. With ones and zeros, we can examine the general structure on the same scale of precision as the diagram provides. But in some cases, we can supply specific numeric values in the matrix that represent the *magnitude* of the effect, as well.

How are these values determined? Because the matrix represents the *program*, it should somehow tell how each part of the aggregate affects each other part in *one period of time*. If it can do this, then one application of the program to the state vector will carry the state vector one step forward in time. Thus, each value has to indicate *how much effect* each part has on each other in *one time period*.

In the special but very frequently used case of *linear equations*, the amount each aggregate affects each other is obtained merely by *multiplying* the affecting aggregate by a *constant value*. In the library circulation system, a certain constant percentage of the reserved books would go into circulation during each time period, and some other percentage would go back on the reserve shelves. We would have to obtain percentages for shelved books that go to the bindery, books checked out to students that are returned to the vault, and so on, for each possible pair of categories of books. If there are eight categories, then the matrix will contain $8 \times 8 = 64$ percentages, each showing the constant fraction of one category that changes into another each time period.

For instance, we may examine the library records and determine that 3 percent of reserved books are checked out in any given week, but only .002 percent of the books in the stacks are checked out each week. In the *row* for "to circulation," the entry under the *column* "from reserved" would be .03, or 3 percent. Elsewhere in the same row ("to circulation"), under the column representing "from stacks," we would place .00002, or .002 percent (see Figure 6.3).

The format of each row of this matrix is like that of an equation reading

$$N_c = \ldots + 3\% \text{ of } N_r \ldots + .002\% \text{ of } N_s$$

where c stands for circulation, r for reserved, and s for stacks. Completing the other column positions in the row would complete the equation and give the entire set of ways that N_c could be affected.

The numbers .03 (3 percent) and .00002 (.002 percent) in this case are

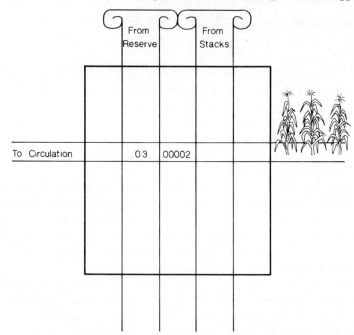

Figure 6.3. How values are placed in the structure matrix representing the percentage of one aggregate (column) that is transformed into another (row). (The columns are the ones with capitals. The row is the one with the row of corn plants, written like an equation across the page.)

determined by *observation*. We would similarly have to observe how many of the N_c were *not returned* during a typical week, since those *remain* as part of N_c for the next week. Suppose that we observe that 21% of the books out at the beginning of any week are returned by the end of that week. Then part of the equation would be the subtraction of those 21 percent, as in the term

$$- .21 \times N_c \ (21\% \ of \ N_c)$$

In the matrix, where would this .21 be found?

The 21 percent are books that went *from circulation* to some other status, so the .21 should go in the "from circulation" column. What *row* should it go in? That depends on the statuses to which the books were returned. Part would go back to *stacks*, part would go to *reserve*, and perhaps part would go to other rows. But we're still missing one element of our fundamental aggregate equation,

$$N_{t+1} = N_t + B_t - D_t$$

The .03 and the .0002 represent "births" into the N_c, and the .21 represents the percentage of the N_c that "died" by being returned. What's missing is the original N_c that were neither born during the week nor died during the week but just stayed where they were, in circulation.

The actual formula for N_c would thus look like this:

$$N_c = N_c - .21 \times N_c + \ldots + .03 \times N_r \ldots + .00002 \times N_s$$

We can combine the first two parts, since they both refer to N_c giving

$$N_c = .79 \times N_c \ldots + .03 \times N_r \ldots + .00002 \times N_s$$

Notice that this is a *computing formula*, a program, for getting a *new* N_c from the N_c and all the other values from *one week previous*. Thus, each time we compute this formula, we move the simulation of the library forward one *week*.

But suppose we didn't want to go forward one week at a time. Suppose, instead, we wanted to go one month at a time, or one day? Does it matter to the values in the formula, and thus in the matrix?

Obviously, the answer is yes, since the values were obtained by observing the circulation pattern for one week. If .21 of the circulating books are returned in one week, then about .03 are returned in one day and about .84 are returned in a four-week month. Well, not exactly, only approximately, but we'll ignore this for the moment. If the approximation were adequate, then the program (formula) for advancing one month at a time would look like this:

$$N_c = .16 \times N_c + \ldots + .12 \times N_r + \ldots + .0008 \times N_s$$

The program for advancing one day at a time would look like this:

$$N_c = .97 \times N_c + \ldots + .005 \times N_r + \ldots + .00007 \times N_s$$

Let's try some actual values in these formulas, to get an idea of the implications of stepping through by different time intervals. Suppose there are 1,000,000 books in the stacks and 10,000 books on reserve. At the start of our study, suppose there are 5000 books in circulation. Then the one month computation would yield

$$N_c = .16 \times 5,000 + .12 \times 10,000 + .0008 \times 1,000,000$$

or

$$N_c = 800 + 1,200 + 800 = 2,800$$

In other words, if there were 5000 books in circulation at the beginning of the month, this model (ignoring other statuses) predicts 2800 books in circulation at the end of the month.

Now let's look at the daily model, which gives

$$N_c = .97 \times 5{,}000 + .005 \times 10{,}000 + .00007 \times 1{,}000{,}000$$

or

$$N_c = 4{,}850 + 50 + 70 = 4{,}900$$

In other words, over a *short* interval of time, the principal effect on N_c is N_c itself, but over a *long* interval of time, other factors become more important. Tomorrow will be pretty much like today, but next month, who knows?

In the previous section, we saw how the diagram of immediate effects was related to the structure matrix. That seemed a clear enough concept, but putting actual numbers down has raised a serious question about the relationship between the two forms. Over a short time interval, the main influence on one of the aggregates is that aggregate *itself*. Simply by sitting there, it "affects" its future value. Therefore, on a short time scale, the system would appear to be "continuous"—to move smoothly from present values to future values.

If we lengthen the time interval, however, there will be little relationship between the N_c at one end and the N_c at the other. In the actual library, we might count the number of books in circulation at the end of each month and get a sequence of values such as 5000, 2800, 500, 8000, 5200, 3100. The values seem to jump all around, and we might easily label this a "discrete" system. If we were asked, "Does the number of books in circulation have much effect on the number in circulation at the next time interval?" we would reply, "No, not much at all."

How then, should we draw the diagram of immediate effects? Should it reflect the behavior of the system over a short time, in which case each box does have an immediate effect on itself? Or should we draw it over a long interval, in which case each box would have little or no effect on itself? There's no right answer, for it depends on what *time scale* the diagram represents. We can make serious errors reasoning from diagrams if we forget that a diagram represents a time scale, not just a "static, pure structure." A "constant lover" over three days may be very fickle over a month.

In continuous systems, there is *always* an arrow of effects from each aggregate's box to itself. That's what we *mean* by continuity—that the system after a small interval of time looks mostly like it did before the interval. It affects itself more than anything else affects it. But since the arrow is always there, we might just as well leave it off. Everyone will understand that it is there.

We must be aware of such conventional omissions if we want to transcribe the digraph correctly into the structure matrix. Which elements of the matrix correspond to the omitted lines *to* each box *from* itself? Obviously, the line *to* box x *from* box x is M_{xx}, for any x. These are the positions along the "main diagonal" or, simply, the "diagonal" of the matrix, the positions marked with ones:

$$
\begin{matrix}
1 & 0 & 0 & 0 & 0 \\
0 & 1 & 0 & 0 & 0 \\
0 & 0 & 1 & 0 & 0 \\
0 & 0 & 0 & 1 & 0 \\
0 & 0 & 0 & 0 & 1
\end{matrix}
$$

If we are modeling a continuous system, we shall therefore have to insert these diagonal elements in our structure matrix.

In discrete-time systems, on the other hand, the arrow to x from x need not exist, so there can be no convention about omitting lines. In such systems, the convention is to draw *all* arrows that do exist, since continuity from one time period to the next cannot be assumed. For example, when we interpreted Figure 5.2 as a model of a junior high school, we stated that some students might be kept in a grade from one year to the next. Following the continuous convention, we did not place those arrows, but in another situation they would be essential. Suppose, for instance, that the school has a policy of never flunking anyone in seventh or eighth grades, but a student may be held from graduating from the ninth grade. Then with a time step of one year, there would be no line from box 1 to itself (seventh grade), but there must be a line from box 3 to itself (ninth grade), to represent the possibility of a student being held back.

The matrix representing this school policy would be

$$
\begin{array}{c|ccccc}
 & S & 7 & 8 & 9 & K \\
\hline
S & 0 & 0 & 0 & 0 & 0 \\
7 & 1 & 0 & 0 & 0 & 0 \\
8 & 0 & 1 & 0 & 0 & 0 \\
9 & 0 & 1 & 1 & 1^* & 0 \\
K & 0 & 1 & 1 & 1 & 0
\end{array}
$$

with the only change being the placement of the starred element in the diagonal position M_{44}, representing a promotion to ninth grade *from* ninth grade—or, in plain language, a flunk. We now see that the original matrix should have been

$$
\begin{array}{ccccc}
1^* & 0 & 0 & 0 & 0 \\
1 & 1^* & 0 & 0 & 0 \\
0 & 1 & 1^* & 0 & 0 \\
0 & 1 & 1 & 1^* & 0 \\
0 & 1 & 1 & 1 & 1^*
\end{array}
$$

with the added ones on the diagonal indicating the possibility of flunking *any* grade.

Previously, we recognized a *source*, because it had all zeros in its row—that is, all its "from" positions were empty, and nothing had an influence on it. Now, we might want to think of the source as influencing itself, so it could have a one in the diagonal position of its row. Thus, whenever we find row s with a one in column s and nowhere else, we recognize a source, s. In the school model, the source row would be

$$
\begin{array}{ccccc}
1 & 0 & 0 & 0 & 0
\end{array}
$$

The interpretation of $M_{11} = 1$ is that students may be held from entering the school, if they are deemed not ready for junior high.

Previously, we recognized a *sink* because its "from" column was all zeros. Now, there could be a one in M_{kk}, indicating that once graduated, they *stay* graduated. If we use this approach in the matrix, we'll find that the sink accumulates all of the individuals that have passed out of the system. This may be precisely the value we're interested in, rather than the internal workings that led to this excretion.

Projecting Possible Effects with a Structure Matrix

. . . and if necessary trace back the sources of its movement through all its successive stages to the original moving power . . . being convinced from experience of the vast power which analysis derives from the great condensation of meaning in the language it employs, I was not long in deciding that the most favourable path to pursue was to have recourse to the language of signs. (Babbage 1826:346)

Once we have settled the question of ones or zeros on the diagonal of the effects matrix, we can use a computer to calculate our matrix of possible effects—in strict analogy to our graphic procedure from drawing the diagram of possible effects. This is done by an explicit computing procedure called "matrix multiplication." Actually, in this case, we'll be doing a special kind of matrix multiplication called "boolean matrix multiplication," but it's of no concern to the reader.

Rather than trouble the readers who don't know how to do it, or bore the readers who do, we shall simply relegate the task to our computer. To multiply two matrices, R and S, we simply instruct our computer to do

$$M = R \times S$$

and we can assume that it will know what we mean. For those who want to find out for themselves, any text on matrix algebra, or linear algebra, will teach them this and other tricks (Noble 1969).

On the other hand, we do want to see the *result* of this boolean matrix multiplication. Suppose we take the matrix of the school system and call it M_1 to indicate that it shows the effects through one time step, whatever size step we have chosen in our approximation. That is,

	S	1	2	3	K	
	1	0	0	0	0	S
	1	1	0	0	0	1
$M_1 =$	0	1	1	0	0	2
	0	1	1	1	0	3
	0	1	1	1	1	K

the matrix form of our original bathtub figure. Now if we ask our computer to perform

$$M_2 = M_1 \times M_1$$

that is, to multiply M_1 by itself, the effects, by definition, will be propagated one time step into the future, giving

	S	1	2	3	K	
	1	0	0	0	0	S
	1	1	0	0	0	1
$M_2 =$	1	1	1	0	0	2
	1	1	1	1	0	3
	0	1	1	1	1	K

which might thus be called the "matrix of secondary effects." The zero in the lower left corner, M_{ks}, tells us that the input from s *cannot* reach k in only two time steps, though it can get anywhere else in the system as shown by all the other ones in the s column. For the junior high school, this zero says that there is no way a student can graduate only one year after entry—that is, no way to go "to" k "from" s in one year.

If we now ask our computer to perform

$$M_3 = M_1 \times M_2$$

we project our secondary effects ahead one time period and get the following matrix of tertiary effects:

$$M_3 = \begin{array}{c|ccccc} & S & 1 & 2 & 3 & K \\ \hline & 1 & 0 & 0 & 0 & 0 & S \\ & 1 & 1 & 0 & 0 & 0 & 1 \\ & 1 & 1 & 1 & 0 & 0 & 2 \\ & 1 & 1 & 1 & 1 & 0 & 3 \\ & 1 & 1 & 1 & 1 & 1 & K \end{array}$$

Now there is a one in M_{ks}, which tells us that it *is* possible to graduate within two years of entry. While this conclusion is rather obvious from the graph, on a larger graph this path to graduation might not be so easy to see among the tangle of lines. The matrix form permits us to relegate to the computer the task of finding the fastest path, or "critical path," to graduation, and the computer is not bothered at all by a little more complexity.

As it turns out, if we continue our multiplications past this point, for this system, we would find that $M_4 = M_3$, so that no further changes would take place. M_3 is therefore the matrix of "possible effects," and thus corresponds precisely to the *transitive closure* shown in Figure 6.2. Again, we might have written down the matrix directly from Figure 6.2, but this way we can leave the work to the computer, for in a more complex graph we might miss one of the lines or put in an extra one.

In addition to giving us a simple and reliable way to project the structure of effects or flows or classifications forward in time, the matrix form may be psychologically beneficial in its own right. For instance, the zeros above the diagonal show us that no tub ever has any effect on a lower numbered tub. In other words, the system has no feedback; and not every part of the system can affect every other part. Ink spilled in tub 3 will never darken tub 1.

An example of a system with feedback is shown in Figure 6.4. This system has five tubs, including source and sink. They are connected in the same way as our original bathtub system *except* that the outputs of tub 3 are pumped to tub 1, rather than going into output (K). The matrix of

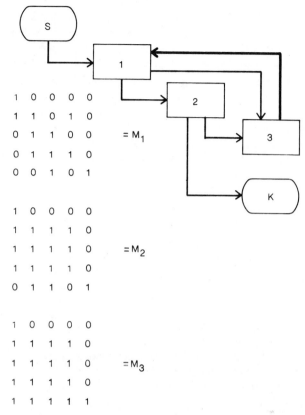

$$
\begin{array}{ccccc}
1 & 0 & 0 & 0 & 0 \\
1 & 1 & 0 & 1 & 0 \\
0 & 1 & 1 & 0 & 0 \\
0 & 1 & 1 & 1 & 0 \\
0 & 0 & 1 & 0 & 1 \\
\end{array} \quad = M_1
$$

$$
\begin{array}{ccccc}
1 & 0 & 0 & 0 & 0 \\
1 & 1 & 1 & 1 & 0 \\
1 & 1 & 1 & 1 & 0 \\
1 & 1 & 1 & 1 & 0 \\
0 & 1 & 1 & 0 & 1 \\
\end{array} \quad = M_2
$$

$$
\begin{array}{ccccc}
1 & 0 & 0 & 0 & 0 \\
1 & 1 & 1 & 1 & 0 \\
1 & 1 & 1 & 1 & 0 \\
1 & 1 & 1 & 1 & 0 \\
1 & 1 & 1 & 1 & 1 \\
\end{array} \quad = M_3
$$

Figure 6.4. A diagram of the immediate effects obtained by adding one feedback path (heavy line) to the original bathtub system. M_1 is the corresponding matrix of immediate effects, M_2 is the matrix of secondary effects, and M_3 is the matrix of tertiary effects. Because $M_3 = M_4$, M_3 is also the matrix of ultimate, or possible, effects.

ultimate effects (M_3) shows that only the source, S, is unaffected by any other tub, and that the sink, K, is the only tub that does not affect any other tub. Other than these two tubs (whose behavior we predict from their *definition* as source and sink), all parts of the system may *possibly* have an effect on one another.

A system in which any part may have an effect on any other part is called "completely connected." A system in which any part either affects or is affected by (or both) every other part is called "connected." Our original bathtub system was connected, but not completely connected. The system of Figure 6.4, because of the addition of one line, *is* completely connected. In plain terms, that means that ink spilled in any tub will pollute all the others after a while.

Sometimes, we find that a system is not even connected, let alone completely connected. In that case, there is at least one part that neither affects nor is affected by some other parts. When this happens, we can consider the disconnected part or parts separately from the rest. In a very real sense, a disconnected system is two (or more) systems, not one.

Partitioning a System

> I shall rot here, with those whom in their day
>   You never knew,
> And alien ones who, ere they chilled to clay,
>   Met not my view,
> Will in your distant grave-place ever neighbour you.
>
> No shad of pinnacle or tree or tower,
>   While earth endures,
> Will fall on my mound and within the hour
>   Steal on to yours;
> One robin never haunt our two green covertures.
>
> Some organ may resound on Sunday noons
>   By where you lie,
> Some other thrill the panes with other tunes
>   Where moulder I;
> No selfsame chords compose our common lullaby.
>
> The simply-cut memorial at my head
>   Perhaps may take
> A rustic form, and that above your bed
>   A stately make;
> No linking symbol show thereon for our tale's sake.
>
> And in the monotonous moils of strained, hard-run
>   Humanity,
> The eternal tie which binds us twain in one
>   No eye will see
> Stretching across the miles that sever you from me.

> Thomas Hardy
> *In Death Divided*

Among people, even the sadness of death itself is surpassed by the utter despair of perpetual separation. We are a gregarious lot. With modern telephony and transportation, we have been reared to feel the connectedness of the world, rather than the separation characterizing our species before this long-distance century. With electromagnetic waves, "linking symbols" regularly go back and forth between Earth and Mars. Separation as complete and hopeless as Hardy's must, for us, be merely a question of economics, or a fiction. But it's a fiction that systems thinkers love to imagine, not out of a sense of morbidity but for purely practical reasons.

Because of the Square Law of Computation, systems thinkers are ordinarily delighted to discover that one system can actually be considered as two or more disconnected ones. On really large systems, however, perhaps with hundreds of elements in the state vector, it's almost impossible to recognize such simplifications. Drawing a picture becomes a nighthaps with hundreds of elements in the state vector, it is almost impossible mare, and the value of alternative representations increases.

Figure 6.5 is an example of a system with seven parts and one input and one output. Below is the matrix, M_1, representing that diagram. The matrix M_p is the matrix of possible effects, derived from M_1 by our standard computer procedure.

In this form, the matrix of possible effects, M_p, is not fully revealing. But we do notice that many of the elements are zero, showing that at least the system is not *completely* connected. Can we follow this clue and discover more about the system's structure?

Suppose that the diagram of this system had originally been drawn with different numbers on the boxes. By the Principle of Indifference, the system would still be the same. Consequently, we may arbitrarily relabel such a diagram without changing the abstract structure it represents. But how do we renumber a matrix?

Consider our highway distance matrix:

	Chicago	New York	Seattle
Chicago	0	840	2064
New York	840	0	2904
Seattle	2064	2904	0

Suppose we decided to call New York "Big Apple." Would that change the distances? Of course not, so whatever we do in realphabetizing our matrix shouldn't change the distances between cities. We may accomplish the change in two steps. First, we interchange and rename the Chicago and New York *columns*, giving

	Big Apple	Chicago	Seattle
Chicago	840	0	2064
New York	0	840	2904
Seattle	2904	2064	0

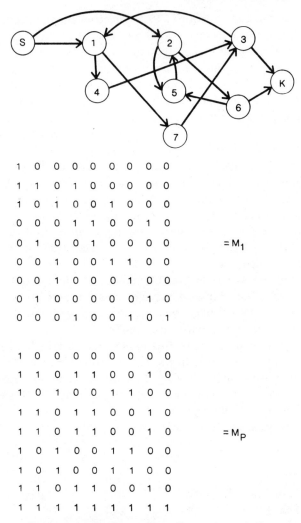

Figure 6.5. A diagram of immediate effects for a system with one input (s), one output (k), and seven parts (1–7). M_1 is the corresponding matrix of immediate effects, and M_p is the matrix of possible effects.

Then we must interchange and rename the Chicago and New York *rows,* to complete the alphabetical ordering, giving

	Big Apple	Chicago	Seattle
Big Apple	0	840	2904
Chicago	840	0	2064
Seattle	2904	2064	0

In precisely the same way, we may renumber the rows and columns of a structure matrix. By the Principle of Indifference, it doesn't matter how often we do this; but by the Principle of Difference, it may matter a whole lot. Thus though the M_p of Figure 6.5 seems just a jumble of ones and zeros, if we interchange numbers 2 and 7, we get an equivalent matrix shown in Figure 6.6. Here, the division of the internal rows and columns (one through seven) would be obvious even without the helping dotted lines. The interior matrix has two distinct parts, neither of which has any ultimate effect on the other. The first part (1, 2, 3, and 4) is represented by the ones in the upper left "submatrix"; whereas the second part (5, 6, and 7) is represented by the 3 by 3 submatrix in the lower right.

If we redraw (and renumber) our original digraph, we get the diagram shown in Figure 6.6. Now the division of the systems into two *independent* subsystems is readily apparent, but would we have thought of redrawing it without seeing the matrix in partitioned form? Clearly, the matrix representation can reveal structure concealed in the complexity and arbitrary arrangement of a structure diagram, and vice versa.

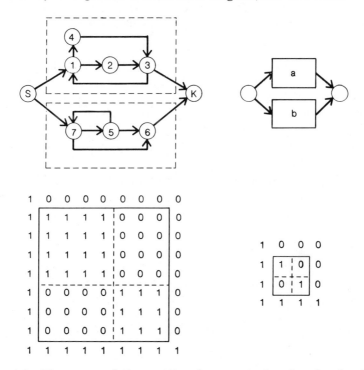

Figure 6.6. The system of Figure 6.5, redrawn, renumbered, and reduced to a simpler system, taking advantage of the partitioning shown in the matrix of possible effects.

The system represented in Figure 6.6 is a disconnected, or separable, system. Often, if we write down the structure matrix of a very large system we discover that a suitable rearrangement of rows and columns yields a form like that of Figure 6.6, in which case we can *reduce* our image of the system to two separate systems, as shown by the parts *a* and *b* in the Figure. Such a reduction helps us to beat the Square Law of Computation.

But in very large systems, it is unlikely that we will ever be able to partition the matrix so cleanly. Sometimes, scattered among a large block of thousands of zeros there will be a handful of ones. Because we are limited in our computational capacity, we may want to try *pretending* that those ones are not there. If the system obeys the Law of Effect, that "small" change will have a "small" effect on the behavior, in which case we say that the system is "almost separable," or "almost disconnected." If Fortune smiles upon us, this ruse will give us a very good approximation to system behavior; if we are a bit less fortunate, it may at least give us a good first approximation; but if we are out of favor with the gods, we will eventually have to face the day of reckoning. Still, if you're afraid to gamble, you'll never be much of a generalist.

QUESTIONS FOR FURTHER RESEARCH

1. *Anatomy and Folk Music*

There is a folk song called "Dry Bones" which relates the tale of Joshua raising up the bones and giving them life. In the song, the famous refrain goes something like ". . . the knee bone's connected to the thigh bone, and the thigh bone's connected to the hip bone . . ."

Write down a structure matrix embodying the anatomical view reflected in this song, or if you cannot remember the song, draw one from your memory of the human body. (Do *not* go to a graveyard for this problem.)

2. *Demography*

In addition to births and deaths in a fixed population, the movements of peoples from place to place can be described by bathtub systems. Migration statistics are available in Hauser and Duncan (1959), as well as descriptions of the problems of demographers. Try to model some of the described population flows, and record the difficulties encountered in the attempt.

3. *Urban Problems*

Beer cans are only one of the aggregates that are beginning to cover the United States. We have aluminum in other forms, but also iron, tin, mercury, lead, cadmium, plus many other metals and nonmetals. Pick one of these substances and trace its transport through the systems in which we live—including our own bodies—using structure diagrams to reveal the sinks in which they are now accumulating. Discuss possible mechanisms that could be created to connect these sinks back into the cycle, so as to eliminate or reduce the accumulation of material in them.

4. *Economics*

What the economist calls "free goods" correspond to the class of sources and sinks, in our terminology. That is, a good is free if no possible use needs to be suppressed so that some other possible use can occur—that is, use does not affect a free good. A source is a free good in the positive sense, while a sink is a place for disposal of "bads," as opposed to the acquisition of goods.

Of course, source and sink are only approximations, as is the concept of free good. Discuss the free goods approximation in economics, including instances where a theoretically free good was in fact limited as new usages were created or the magnitude of old uses grew. In addition, discuss the economic adjustments to those limitations.

5. *Developmental Psychology*

The general idea of *conservation* as a reasoning device is developed in children at a relatively early age. This has been shown most vividly by Piaget and his followers in experiments involving the pouring of liquids among containers of varying shapes and number. Discuss the development of this heuristic from early childhood to the mathematical sophistication of scientifically trained reasoning, including examples such as water-alcohol mixing and trick experimental arrangements in which conservation reasoning does not work. In addition, give examples of the difference in conception between discrete and continuous—extensive and intensive—quantities. (References: Piaget 1952; Bruner 1966).

7

Structure and Behavior

It is, I think, one of the substantial advances of the last decade that we have at last identified the *essentials* of the "machine in general."

Before the essentials could be seen, we had to realize that two factors must be *excluded as irrelevant*. The first is "materiality"—the idea that a machine must be made of actual matter, of the hundred or so existent elements. This is wrong, for examples can readily be given showing that what is essential is whether the system, or angels and ectoplasm, if you please, *behaves* in a law-abiding and machine-like way. Also to be excluded as irrelevant is any reference to energy, for any calculating machine shows that what matters is the *regularity* of the behavior—whether energy is gained or lost, or even created, is simply irrelevant.

The fundamental concept of "machine" proves to have a form that was formulated at least a century ago, but this concept has not, so far as I am aware, ever been used and exploited vigorously. A "machine" is that which behaves in a machine-like way, namely, that its internal state, and the state of its surroundings, defines uniquely the next state it will go to. (Ashby 1968)

Ashby's definition of a machine is noteworthy for its restrained language. He translated the pithy word, "behaves" into cold, but precise, language. In effect, he says:

> *A machine is that which has an internal state and exists in an environment, which also has a state. These two states define uniquely the next internal state, at all times.*

Nowhere is there reference to structure, for Ashby considers structure to be a concept *derived* from behavior. As the states progress onward in time, some aspects of the state may prove, empirically, to be stable. The concept of structure, therefore, derives from the concept of stability, and not vice versa.

This conception of the relationship between structure and behavior is utterly contrary to the most widely held view—namely, that structure *determines* behavior. Such a divergence of view indicates that it will be worth our while to spend a chapter on the age-old dichotomy, structure and behavior.

The Structure of Structure

> O body swayed to music, O brightening glance,
> How can we know the dancer from the dance?
>
> W. B. Yeats
> *Among School Children*

Yeats' disturbing couplet can hardly be considered a matter of daily concern to the practicing scientist or plumber. When presented with the bathtub drawing, each of us immediately and instinctively *knew* how to partition the system into structure—the tubs—and behavior—the water. We knew because our culture knows these things. Other cultures, possibly those more intimately concerned with water, might see it differently. But we *knew*.

What has our culture taught us about structure and behavior? We know that structure is hard, enduring, and cleanly defined in porcelain, like a bathtub. We also know that behavior is soft, evanescent, and rippled even by slight breezes, like water. Yet once we have reached the level of abstraction of the structure matrix, the actual tubs have faded from our memory, along with all our cultural clues. Only then can we finally see that the "structure" of the system need have no connection whatsoever with our culture-bound image of "materiality," or enduring physical parts.

True, in the bathtub example, it is the *enduring* physical parts—the tubs and drains—which are reflected in the structure matrix, and the *variable* physical parts—the water—which are reflected in the population vector. But that which is physical need not endure and that which endures need not be physical. Water is certainly as physical a thing as the iron or porcelain of the tubs. The ocean is largely made of water, yet its level is one of the more stable aspects of our environment. On the other hand, the "tub" containing the ocean is constantly changing. One need only look at the changing map of Holland over the past century. The entire coastline has been distorted by diking and filling without measurably changing the water level of the ocean. So it is only in some particular systems that solid material happens to be enduring and liquid, variable.

Structure need not be physical at all. An enduring idea, for example, such as "freedom of speech and of the press," forms part of the structure of a nation just as does an enduring building. We do not question here that such things as ideas have an ultimate physical form, such as ink on paper, or chemical or electrical states in human brains. Indeed, if we wish to put our systems thinking on a basis of "materiality," we can build up to such "nonmaterial" structures as ideas, social structures, and laws. When we say that structure need not be physical, however, we mean that it need not be what the observer is accustomed to calling physical—namely, hard solid matter.

Consider, for instance, the type of problem with which demographers often deal. We start with some population of people and wish to study how they will multiply or decline in the future. At a very simple level, we might divide the population into three age groups: 0–14, 15–44, and 45–120. To the level of approximation of our model, only the middle group produces babies, and they are introduced—at age zero—into the 0–14 group. Assuming no immigration or emigration, the only other way of affecting the population size is by deaths, which can occur in any group but may be more likely to occur among the oldest.

What we have just described in words can be shown in the digraph or structure matrix of Figure 7.1. The structure matrix is restricted to 3 by 3 because we have no immigration or emigration and we are not interested in counting the dead. This digraph and matrix represent structure in precisely the same way as did the digraph and matrix of the bathtub example, even though the structure it represents is not physical in the usual sense of that word.

For instance, consider the zero in the upper left corner (M_{11} for row 1, column 1) which is a representation of the fact that to a very good approximation, the age group 0–14 does not produce offspring. Although this is partly a biological fact about human beings, it is not entirely so, for most human females are physically capable of producing offspring before they reach the age of 15. Indeed, it would be possible to imagine a culture in which the *norm* of behavior was to impregnate girls as soon as they reached puberty, and to prohibit sexual intercourse after the age of 14. In that society, the zero in M_{11} would be a one, and the one in M_{12} would be a zero. Thus part of the structure of this system represents a belief system, part represents a biological system, and part represents our convention by which one age category is separated from another at a particular age.

To the demographer, then, the structure is the ensemble of relation-

DIAGRAM OF IMMEDIATE EFFECTS

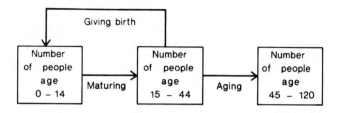

STRUCTURE MATRIX
(15 year intervals)

```
0   1   0
1   1   0
0   1   1
```

Figure 7.1. A directed graph (digraph) and structure matrix representing a group of people divided into three age categories, showing that only people in the 15–44 group can have children, and thus cause an increase in the 0–14 age group. The structure matrix gives the same information in a manipulable form.

ships that are relatively fixed and that determine the changes in the population size. At this level of analysis, she cares not whether this structure derives from biology, culture, or astrology. To her, two populations exhibiting the same structure *are* the same, regardless of the source of that structure. Because of this indifference, she is able to construct mathematical systems—models—with the "same" structure as populations of people. Most of the demographer's time is spent with these models, rather than with people. The advantage of the models, of course, is that she may study systems which she cannot find, or produce, or wait to examine, such as societies which prescribe mating for 11-year-old girls.

To insist on demonstration of "materiality" as the source of structure in a model is to be spellbound by an analogy. Our structure diagrams *look* like physical structures, like interconnected tubs of water or sheep pens. They should. We *designed* them that way, to take advantage of the way our brains have evolved with visual systems in a world of solid "objects." Because our models *look* like sheep pens, we may, just may, find them aiding our intuition. But it's not their *appearance* that's the essence of their structure. It's their *endurance*. Structure is that which stands, which remains, which is unchanged, regardless of its physical properties.

From physics, we have inherited some terminology and an excellent track record. Why not! The physics of mechanical objects is very close to the type of physical world our sensory apparatus evolved in. Other sciences work a bit further from this world of direct perception. If they want to be more like physics in terms of success, they will have to break the mystical spell of "materiality." Otherwise they will resemble physics—but only in terminology, not in success.

Projecting Behavior with a Linear Program

Priests are like mules: they eat a lot, and, though they don't reproduce, there's always enough of them.

Alpine Proverb

The conceptual tools which we use for conservative systems can be extended to nonconservative systems, wherein members may disappear or, like priests, miraculously appear. But when we lose the constraints of conservation, there is less we can say in general about the system's behavior simply from looking at the form of the equations, the digraph, or the structure matrix. To make even gross predictions, we have to replace the simple "ones" in the structure matrix by more precise descriptions of the relationships that exist. In doing so, however, we try to hold the detail to a quantity and type that we can handle with the methods we have at our disposal.

It should be emphasized once again that even in these days of monster computers, we cannot simply write down equations without regard to our ability to solve them. From this methodological trap there are two paths of escape. In the first approach we may write down the equations in unadulterated form and then try to draw general conclusions about the *nature* of the solution in some general terms. This procedure is much in keeping with the general systems approach and has an old, if not well known, tradition in mathematics. For example, in 1881 Henri Poincaré introduced a subject he called "qualitative dynamics" (see Piexto 1967) by which one tries to describe the long-term behavior of a system in qualitative terms without solving the equations.

In the second approach, we attempt to write down equations from the outset which are approximate but which we *know* we can solve, perhaps with the aid of a computer. Each approach has its dangers. In the qualitative approach, we may miss the behavior of interest through the gross nature of our solutions, but in the approximative approach, we may miss it through the gross nature of the equations themselves. Yet each approach has its uses, as we hope to illustrate with the following extended example. We will begin by taking the second quantitative approach.

Suppose we wish to study the behavior of our population of Figure 7.1. Evidently, we shall have to provide some rules for birth and death rates, rules that can be arbitrarily complex, leading to unsolvable equations. Suppose, instead, that we decide to approximate our population with simple *linear* birth and death rules, by which we mean that the death or birth rate is always calculable as some fixed fraction of the existing population in one or more of the groups.

If the system were indeed linear in each of the variables, we could take a time step of 7.5 years and specify the entire structure by replacing the "ones" in the structure matrix by appropriate *numbers,* as shown in Figure 7.2. In that matrix, for example, the .4 says that four-tenths of the 0–14 children are still alive and still less than 15 at the end of the 7.5-year period. Of the other children, .1 have died and do not show in the matrix, and .5 have become over 15 and are thus moved into the 15–44 bracket. This .5 is M_{21}—"*to* row 2, *from* column 1," "*to* adults *from* children"— while the .5 in M_{12} represents the "feedback" from the young adult group to the children group—the actual births in the period. In the young adult group, .7 are still young adults at the end of the period, .25 have become old adults, and .05 have died and are not represented in the matrix. The old adults do not reproduce ($M_{13} = 0$), and .2 of them die during the period, leaving .8 in the group at the end (M_{33}).

Given that there is no immigration or emigration in this model, these figures in the matrix suffice to permit our computer to project any initial population vector of a linear model forward in time by 7.5 years at a crack, using the process of "matrix by vector multiplication." The program to do this might say:

1. Repeat lines 2–3 indefinitely, stepping t from 0 by 7.5 years.
2. $v = M \times v.$
3. Plot v versus t.

The first few population vectors produced starting with (200, 400, 900) are shown in Figure 7.2, and the curves for more periods are shown in the chronological graph of Figure 7.3.

There is nothing mysterious about the matrix multiplication. It is merely a shorthand for the process of taking each discriminated aggregate and computing the contributions to it from each of the others, taking away the percentage of its own initial population that is lost. The first digraph shown in Figure 7.2 is the "initial" population of (200, 400, 900). We don't know where it came from. It's just a "given" in this simulation.

LABELED DIGRAPH

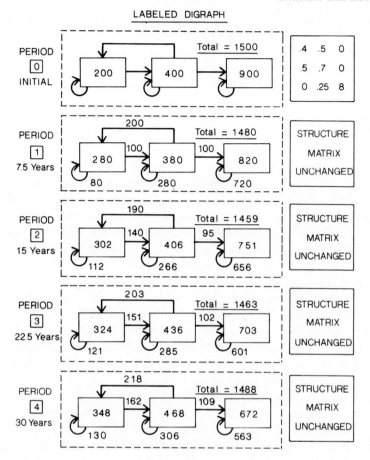

Figure 7.2. The behavior of a three-component population with linear structure worked out in detail for several periods, to show the method by which the present state is projected forward "linearly" at each time step. The numbers on the arrows show where each part of the component came from. For instance, in Period 1, 80 of the children are children who were 0–7.5 at period 0, and didn't die. The other 200 were born to mothers in the 15–44 age group.

We could just as well have started with some other vector, and we might well do that in trying to understand the structure of this model more fully.

The second digraph, however, is *not* arbitrary, once we accept the initial condition as given. Each of the values in (280, 380, 820) is completely determined by applying the matrix of the system's structure to the state vector of the previous state. "Applying" here means executing the program, which means matrix-by-vector multiplication. The number of children, v_1, is determined by applying the first row of the matrix, (.4, .5, 0),

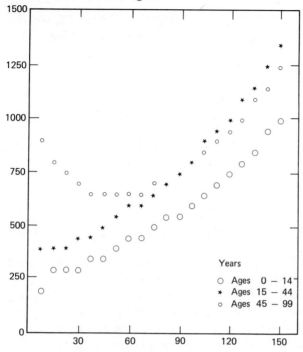

Figure 7.3. A computer plotted simulation of the growth of the three population components, starting from a population of (200, 400, 900) and projected forward by successive applications of the matrix of the system's structure.

to the vector (200, 400, 900), multiplying each element by the corresponding element, and adding them together, as in

$$.4 \times 200 + .5 \times 400 + 0 \times 900$$
$$80 \quad + \quad 200 \quad + \quad 0 \quad = 280$$

Of the 280 children, 80 represent .4 of the original 200 who were 0–7 years old at the initial time, and who thus are now 7.5–14.5 years old. The other 200 represent the .5 birth rate applied to the 400 young adults in the previous 7.5 years. That is, on the average, this model says that each two adults produce one child every 7.5 years. There are no children derived from the "old adults," even though there were 900 of them initially. Old adults do not, in this model, reproduce, and they certainly don't rejuvenate into 0–14 year olds.

Each of the other two components, v_2 and v_3, are computed in a similar manner, by applying rows 2 and 3 to the initial population vector to bring in the contribution of each part. In Figure 7.2, we also show the *total* population, N, which is obtained by simply adding

2.1 $N = v_1 + v_2 + v_3$

which our computer can do very nicely.

Our computer can also produce a graph, shown in Figure 7.4, by adding one more step to its program:

3.1. Graph the value of N versus the current value of t.

and the graphs of Figure 7.3 by adding another:

3.2 Graph the values of v_1, v_2, and v_3 versus the current value of t.

Taking full advantage of our computer in this way, *quantitative* reasoning about the system comes well within the grasp of the mathematical novice. If these curves do not track our data on the actual behavior of the system, we can change some part of the model and try again. To be sure, when the model *does* track the system accurately, we can never be positive that the model's veracity can be extended beyond the observed data. We can't even be positive that it's a reliable map of the present structure of the system. Therefore, we mustn't jump on the bandwagon of the quantitative model just because it doesn't look obviously wrong. It's better to be skeptical, even after we've published our results.

Qualitative Reasoning About Structure

The controversy between the followers of the physics of Descartes and of Newton was at its height at the end of the Seventeenth century. Descartes, with his vortices, his hooked atoms, and the like, explained everything and calculated nothing. Newton, with the Inverse Square Law of gravitation, calculated everything and explained nothing. History has endorsed Newton . . . But I am certain that the human mind would not be fully satisfied with a universe in which all phenomena were governed by a mathematical process that was coherent but totally abstract. (Thom 1975)

In order to acquire some preliminary insight into the meaning of structure, we're now going to turn our simulation of Figures 7.2 to 7.4 around. It started as a "white box" simulation. That is, we had full access to its entire internal structure, as well as its initial state. This is the way the

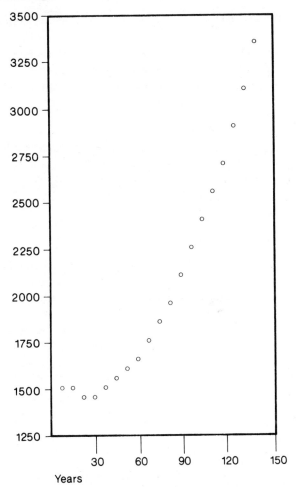

Figure 7.4. A computer plotted simulation of the growth of the total population (the sum of the three components of Figure 7.3). The slight dip at the beginning shows that the growth is not purely exponential.

anatomist knows structure, but now we want to take the ethologist's view, or at least the physiologist's. We want to observe, for the moment, *only the behavior*. This is the "black box," whose insides are completely closed—or so we'll pretend. In fact, they'll be *so* closed that we won't even see Figure 7.3, but only the total population growth of Figure 7.4.

The most striking aspect of the growth in Figure 7.4 is that it seems to be *exponential*. Yet if we look a bit more carefully, we can see a tiny dip, just at the beginning. In real life, we might first wish to dismiss this dip as an observational error, because without it, our model could be vastly simpler.

Keep in mind that we're making a black box observation of the total population, N. We're entirely ignorant of the age structure we put in the white box. Thus our simplest model would involve a single bathtub equation of the form

$$N' = f(N)$$

In this form, there's no input and only one type of individual. We don't know what function f represents, but since the curve *looks* rather exponential, we might guess that

$$N' = pN$$

where p characterizes the *rate* of growth, the proportionality factor. In other words, the behavior of the population in this model doesn't depend on anything but the total size of the population itself. Can this model explain the observed behavior in Figure 7.4?

Here we may try some qualitative reasoning. Recall that N' is the *slope* of the chronological graph of N. Therefore, if N' depends *only* on N, whenever N has a certain value, the slope must have the same value. But look at Figure 7.4. At the beginning, N is 1500 and sloping *downward*, while a little later it is 1500 and sloping *upward*. Therefore, we must conclude that

$$N' = f(N, \ldots)$$

A population whose growth rate depends *only* on its size cannot show dips or bumps—it can only rise or fall monotonically or stay exactly the same. We must, therefore, look for something else in the system besides N, if we want to explain the dip we observe. Although *we* can see, as in Figure 7.2, that the "something else" is the age structure of the population, the black box observer doesn't know that it is there and must search for an explanation.

Perhaps the explanation is that there has been immigration or emigration? Suppose we know from our records that the system is entirely isolated, as, for example, a population on an island might be. Then we are forced to conclude that *within* the system there is something more than a simple dependence on total population. That "something more" might just be an arbitrary change of the rules—"things just changed"—but, as scientists, we would not be too happy with that explanation. What we would like to find is a *fixed* set of rules. But then the "something else" would have to be reflected in the rules themselves—and that "something else" is what we call "structure."

Structure, in other words, is "that which stands"—that which remains uninfluenced by change coming from outside of a system. In terms of our fundamental system equation

$$S_{t+1} = F(S_t, I_t)$$

a perfectly *structureless* system would have

$$S_{t+1} = I_t$$

Such a system simply follows the input. It has no "mind of its own." The temperature of a piece of metal simply rises and falls as the room temperature changes—almost structureless, but not quite, since the rise or fall is not instantaneous.

A computer memory cell is a kind of structureless system—when we write into it, the previous state is immediately and completely forgotten. This feature is what makes the computer so "protean," though Proteus is perhaps not the correct model of a structureless system. Proteus was a sea god and changed his shape when seized to avoid capture. He did not assume the shape of his captor, but rather a shape to *foil* his captor—hardly structureless behavior.

The computer memory cell is not actually structureless either, for it does retain its contents when there is no input. It has the structure of *continuity*—maintenance of state under null input—though it is structureless under the influence of an explicit input. Were it otherwise, it would not be a "memory" cell, and would be of no value whatsoever.

Our black box definitely doesn't follow its input, if there is no input. What kind of reasoning could we use to explain the dip before the rise of Figure 7.4? After assuring ourselves by the most careful search that no members of the population were "leaking" out or "sneaking" in, we could eliminate the I_t from the general system equation, giving

$$S_{t+1} = F(S_t)$$

or in the continuous form

$$N' = f(N)$$

We don't know the exact nature of F, or f, but our previous reasoning told us that *no* form of functional dependence involving *only* S_t, or N, could give us the observed dip, so we can rule out some complicated functional form.

What does that leave? Observational error is always possible, so we check our procedures again, but still find no cause to disbelieve the numbers. Alas, we are forced to conclude that either we have made an error in getting to this point, or else we have misjudged the simplicity of the state of the system. If we've made no error, the state must have more than one discriminated aggregate, so that there is more than one program, or equation, producing Figure 7.4.

Because we were merely pretending to have a black box, we know that this is the correct conclusion. Be careful, though, that you don't forget all the possibilities we had to eliminate before we could reach this conclusion definitively. Only then can we proceed with confidence to further steps that will determine the nature of the system's state.

Once we've decided to allow more than one aggregate in the state, many different explanations offer themselves. Because the program is fixed, however, all the explanations boil down to some effect of the *initial sizes* of the discriminated aggregates. In the simplest case, there would be two aggregates, one which was *decaying* exponentially and the other which was *growing* exponentially. The dip at the beginning is caused by the decaying aggregate being initially larger than the growing one.

We know that the model is somewhat different than this simplest case, but not terribly different. The initial dip in our model *was* caused mainly by the substantial decay of v_3, the old people, who represented 900 of the original 1500 members. No amount of black box reasoning could have told us *exactly* what the initial state vector was, but we can see that we've managed to establish a firm foundation for further research into the question. And that's what general systems thinking is all about.

Detecting the Presence of Structure

I offer this work as the mathematical principles of philosophy, for the whole burden of philosophy seems to consist in this—from the phenomena of motions to investigate the forces of nature, and then from these forces to demonstrate the other phenomena.

                                                    Isaac Newton
                                                    from the preface to *Principia*

Because a *structureless* system simply *follows the input*, we can detect the presence of structure by noticing when a system does *not* follow its input. If we plot the input and the system's response to that input on the same scale on the same chronological graph, the two curves will lie on top of one another if there is no structure. Where they differ, then, we detect the presence of structure.

In Figure 7.5, we see the results of an experiment in which the temperature at the center of an iron bar is measured against room temperature. When the room temperature is raised sharply, the bar temperature does not simply follow, but displays a lagged ogival rise. To account for the shape of this rise, the physicist must consider the physical structure of the bar.

If we repeat the experiment with an ironworker instead of an iron bar, we see even more evidence of structure, as shown in Figure 7.6. If we have the ironworker swallow a transmitting thermometer and we monitor it as the blast furnace heats the air, we see that there is no appreciable rise in his internal temperature at all. And even when a blast of cool air drops the temperature suddenly, his body temperature simply continues on its own, slightly fluctuating, course. Clearly there is structure in this man of iron.

The fundamental laws of *mechanics* are laws about how things behave in the absence of structure, or at least with *most of the structure re-*

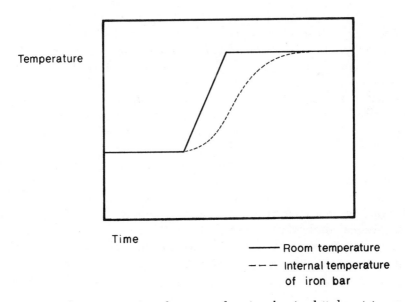

Figure 7.5. The temperature at the center of an iron bar is plotted next to room temperature. To the extent that the internal temperature (the "output") does not follow the room temperature (the "input"), we know that there must be some "structure" in the bar.

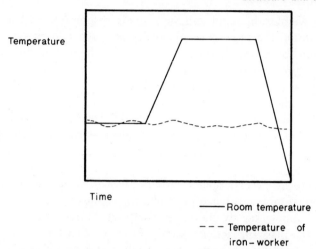

Temperature

Time

———— Room temperature

– – – Temperature of
iron – worker

Figure 7.6. The temperature in an ironworker's stomach shows essentially no relationship to the temperature in the room, even when the blast furnace causes a sudden
rise, or cool air causes a sudden drop. We conclude from these observations that there
must be a great deal of structure interposed between the input (room temperature)
and output (stomach temperature).

moved. Newton's First Law stated that, if undisturbed by external forces
(inputs), a body would continue in uniform motion. This was important
because it provided the baseline against which change could be detected.

From one point of view, uniform motion could easily have been
thought to be a change, but Newton asserted that uniform motion was
constancy. That is, instead of focussing on *position* as the state variable,
Newton turned his attention to *velocity*, v, the rate at which position is
changing. If $v' = 0$, Newton said, then there is no need to look for the
presence of inputs (unbalanced forces), even though it looks to us as if
the system is changing, for it requires no inputs at all to continue moving
with constant velocity in a straight line.

When we see a deviation from this constant velocity, we infer the
presence of unbalanced force, F. Or, if we *impose* a force F on the system, we may observe v', the rate of change of v, to infer the structure of
the system. To make this inference, we use Newton's equation relating
force (F), mass (m), and acceleration (a):

$$F = ma$$

In our terms, since acceleration is the *rate of change of velocity,*

$$F = mv'$$

Solving this for m, we get

$$m = F/v'$$

This equation tells us that the mass of a simple physical system is its only *structure*, for m is measured by comparing the input, F, to the response, v'. This is exactly what we did when comparing external temperatures to internal temperatures to detect structure in iron bars or iron-workers. We know that there is a large mass—that is, much structure—when a large force gives only a small change in velocity. Anyone who has tried to push a large boat away from a dock will understand the meaning of mass as resistance to change in velocity.

Newton's equation applies only to certain "simple" bodies, namely those whose structure *can* be completely characterized by one number, the mass. Most real bodies have other structure, as well, so they don't respond to forces simply by moving away in a straight line with only the speed of motion to be determined by the structure, or mass. A *real* boat might not move *straight* away from the dock, even slowly, for a variety of reasons, each of which the physicist interprets as additional structure. When you push on the hull, the hull may simply bend a little, or dent permanently, rather than the entire boat moving away. In that case, the physicist speaks of "internal" structure, or "elasticity." Or the boat may move in a semicircle, if it is tethered at one end, rather than in a straight line. Here the physicist speaks of "external constraints," or postulates other "forces" acting through the rope. Or perhaps the boat responds by saying, "Ouch, don't push so hard!" In that case, the physicist won't have much to say at all, except that this is "not a physical situation."

When the object does *not* respond with acceleration in the same direction as, and in proportion to, the applied force, the physicist knows she's dealing with a system with more structure than the simplest of all physical systems, the so-called "mass point." For example, suppose she enters a completely darkened room in which the only visible object is a luminous ball suspended in the center. Is it a mass point? She can't know simply by looking at it. To find out, she must *do* something. She must apply some input force and compare it with the observed response.

If she gives it a tiny push, yet it *does not* move in a straight line, but swings back and forth instead, she will conclude that "there is some *other* force acting on the ball." If the lights are raised and she sees that the luminous ball is a light pull on the end of a string, she may now attribute that "unbalanced force" to the "structure" of the string. With the lights out, however, the Principle of Indeterminability tells us that she cannot say for certain whether "there are unbalanced forces" or "there is structure—the string."

The physicist knows she has a mass point only by performing experiments—by pushing the luminous ball in various directions. Each different push is calculated to reveal a different kind of structure. The luminous ball must have some structure to remain identified as a luminous ball. We cannot, indeed, even *detect* the presence of a system which doesn't have *some* structure. We would walk around in a dark room and bump into this structureless "object" without even noticing, for it would simply yield to every bump. *Failing to yield* is the only clue we have to the existence of structure, for by the Principle of Invariance, we can only understand permanence through attempts at change.

QUESTIONS FOR FURTHER RESEARCH

1. *Unidentified Flying Objects*

Suppose an unidentified flying object lands on your front lawn and just sits there, doing nothing. How would you go about investigating its structure? How would a physicist do it? A chemist? A psychiatrist? A military strategist? A diplomat? A farmer? A dog?

2. *Art History*

Once we withdraw the paralyzing venom of "materiality" from our concept of "structure," the parallel between art and science is seen in much more clarity. Science has its esthetic and art has its regularities. Moreover, the two have not been quite so independent over the ages as we might imagine from modern segmented society. Leonardo, of course, was painter and engineer, but this combination is too easily dismissed by scoffers as an accident of genius. But the interrelationships between art and science through their ideas of form, or structure, have not been restricted to this solitary genius. What other connections have existed between the study of form in art and science through history? What value could such connections have in the future? How could they be encouraged? (Reference: Ritterbush 1968).

3. *Sociology*

Human populations can be studied not just by age groups, but by any classification which sociologists can make. For instance, if we can identify social "classes," we can model the society with a structured aggregate in which each component is the number in that class—professional, blue collar, white collar, capitalist, or what have you. The "social structure" of such a society could be the population vector itself, or it could be the matrix showing the mobility among social classes. Show how the structure matrix might look if the social mobility were linear and:

Case A. The society follows an unbreakable caste system.

Case B. A small amount of mobility can take place but only between "neighboring" classes.

Case C. The society is very fluid, permitting movement from one class to another.

4. *Demography*

Our little three-element population probably seems strange to those trained in the methods of demography (Keyfitz 1968; Rogers 1975), because the three age categories are not equal spans of time. The common practice is to divide the population into *cohorts*, groups of people of the same age, with the same interval spanning each cohort. With human populations, this interval is almost always 1 year, though sometimes in considering the global population, a 5-year interval is used.

Discuss the advantages and disadvantages of using a fixed interval for all cohorts. Discuss the relationship between this interval and the time step used in projecting the population forward. What considerations should determine the size of the interval for a given population, assuming it is fixed? Try to imagine biological situations in which it would benefit the simulation to have different sized intervals, as in our division into (prepuberty, fertile, postfertile).

5. *Demography*

The United States Census for 1950–1970 is broken down as follows:

	Under 5	5–14	15–19	20–24	25–34	35–44	45–54	55–64	65+
1950	16	24	11	12	24	22	18	13	12
1960	20	35	13	11	23	24	20	16	17
1970	17	40	19	17	25	23	23	18	20

Make a rough linear model based on the 1960 and 1950 figures, and use it to project forward to 1970 figures. Try to acount for any discrepancies. Make a rough linear model based on the 1970 and 1960 figures and project it forward to 1980, 1990, and 2000. Discuss the value of these predictions.

6. *Anthropology*

Assume that culture is a program generating certain social and intellectual capacities that are uniquely human. Language, rules of behavior, and other symbol-dependent activities are fundamental to the program. These are part of the structure matrix of the system.

We take this structure and its associated behavior vectors so much for granted that we have difficulty explaining human behavior that is *not* culturally programmed—for example, the behavior of the mentally retarded (MacAndrew and Edgerton 1964), of the deaf-mute (Keller 1903), and of so-called "feral" children (Malson 1975). Consider how this intellectual dilemma might be approached by using the following general systems concepts: white box analysis, black box analysis, qualitative reasoning, quantitative reasoning.

7. *Linear Models*

Although linear models of structure may be taken as a kind of base line, or reference, that does not imply that they are so simple that they can be used without caution. For example, we can get very quickly into trouble in a model, such as our population model, when the age classes are not of equal length or are not equal to the length of the time step, particularly if the rate of change is fast, so the linear approximation is poor. For an example of the kinds of pitfalls in linear modeling, see Bosch (1971) on the response of redwoods to various lumbering policies, and the debunking of this article by three letters in *Science* (22 October 1971). Using these criticisms as a basis, discuss the limitations of our simple three-age population model, and how it might be refined to overcome such problems.

8

The Structure-Regulation Law

I met a traveller from an antique land
Who said: Two vast and trunkless legs of stone
Stand in the desert . . . Near them, on the sand,
Half sunk, a shattered visage lies, whose frown,
And wrinkled lip, and sneer of cold command,
Tell that its sculptor well those passions read
Which yet survive, stamped on these lifeless things,
The hand that mocked them, and the heart that fed:
And on the pedestal these words appear:
"My name is Ozymandias, king of kings:
Look on my works, yet Mighty, and despair!"
Nothing beside remains. Round the decay
Of that colossal wreck, boundless and bare
The lone and level sands stretch far away.

Percy Bysshe Shelley
from "Ozymandias"

Ye Mighty, looking on the works of Ozymandias, might despair because they were so great, as Ozymandias thought, or because they demonstrated how ye Mighty in the past had come to a common fate. "Soft" scientists, looking on the linear models of the "hard" scientists, might despair because they were so great, as the hard scientists thought, or. . . . Or what? Before yielding to despair, read on.

The Equivalence of Structure and Input

Here lies the essential point; from her scientific preparation, the teacher must bring not only the capacity, but the desire to observe natural phenomena. In our system, she must become a passive, much more than an active, influence, and her passivity shall be composed of anxious scientific curiosity, and of absolute respect for the phenomenon which she wishes to observe. The teacher must understand and feel her position of observer; the activity must lie in the phenomenon. (Montessori 1912)

If we are to apply the Montessori method, or any method based upon inferring natural phenomena from passive observation, we must proceed in successive stages. Each stage will reveal another aspect, or kind, of structure. The process is complex, especially when the phenomena are as complex as the behaviors of young children. Perhaps we can learn about the limitations of the method by starting with something simpler and less controversial than the education of the young. Let us take a vacation.

On warm summer days near a dairy farm, one of the least controversial activities is swatting flies. As we discussed earlier, constant efforts to eradicate flies on a screened porch *should* produce a constant rate of diminishing the logarithm of the population, as we saw in the logarithmic plot of the exponential. This predicted constancy assumes that the fly population consists of identical individuals, so that only a single number, the half-life, is needed to characterize the entire structure of the aggregate of flies.

Suppose we continue in our "constant effort at eradication" and see the behavior of Figure 3.11, the curve of two exponentially decaying aggregates mixed together. We would probably conclude that there is more structure in the population than we had anticipated, though of course it might be our "constant effort" that has changed. We would characterize this structure by saying that there are two types of fly—one with half-life R and one with half-life U.

Other explanations—isomorphic to this two-element system—are possible, as we saw previously. We chose exponential decay as somehow the "simplest" explanation. Like the physicist's concept of "uniform motion in a straight line," exponential growth or decay is one kind of benchmark against which we can measure an aggregate system for the presence of structure.

An aggregate with a single type of member will have a population vector of one element and a structure matrix of one row and one column. In other words, a single number suffices to describe how the aggregate affects itself, just as a single mass described the behavior of a simple object for the physicist. When the matrix is one-by-one, matrix-by-vector multiplication is just *ordinary* multiplication. If the time step is one half-life, the state transition equation is

$$S_{t+1} = .5 \times S_t$$

which, of course, gives the rate of exponential decay of S.

Where there are *two independent* exponentially decaying components, the population vector has two elements, and the structure matrix is two-by-two. For example, M might be

$$\begin{matrix} .5 & 0 \\ 0 & .9 \end{matrix}$$

The zeros indicate independence—that neither component affects the decay of the other—and the .5 and the .9 give the rates of decay. This population is represented in Figure 8.1.

Starting with an initial population of (1000, 100), in the next time step

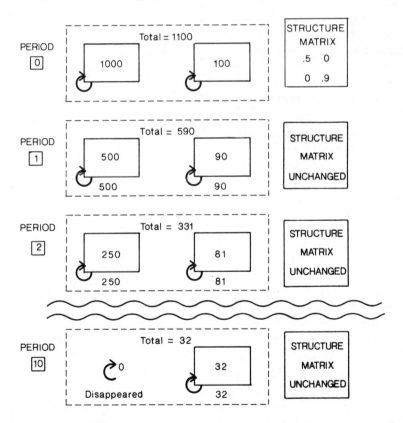

Figure 8.1. The detailed behavior of two independently decaying aggregates. Each aggregate is exponential, and after a short time, the faster decaying one disappears, leaving only a single component, as illustrated in Figure 3.11.

the population would be (500, 90). In successive time steps, the population would become

$$(125, 73) \quad (62, 66) \quad (31, 59) \quad (16, 53) \quad (8, 48) \quad (2, 39)$$

The first subaggregate, originally larger, has decayed faster, leaving only the second, just the behavior illustrated in Figure 3.11. The curves in that figure could be used to reason backward to the presence of two components and to the exact form of the structure matrix.

Other deviations from pure exponential decay can be revealed by more complex behavior of the overall aggregate. In Figure 7.2, we saw that when there was more than one component, and when the matrix did not partition into independent parts, the behavior did not simply rise or fall monotonically. In Figure 7.4, the overall population first fell slightly and then rose, as did the population in the 15–44 age bracket in Figure 7.3. This 15–44 age bracket is *open to input* from the 0–14 bracket, so it can display complexities of behavior not due to its "structure" but to its "input"—the original population vector. Once again, this illustrates the complete black box equivalence of structure and input, as noted in the Principle of Indeterminability (Weinberg 1975: 209–216).

We saw from our black box analysis of this three-aggregate system that a two-aggregate system could show similar behavior—an early decline followed by an exponential rise. We now see that the structure matrix for such a system of two independent aggregates would look something like this:

$$\begin{matrix} .9 & 0 \\ 0 & 1.1 \end{matrix}$$

If the initial vector were (900, 600), the first step of the program would give (810, 660), a net drop from 1500 to 1470. The next few steps would give overall populations of 1455, 1454, and 1468—qualitatively similar to the small dip followed by a rise displayed by the three-aggregate system.

If our observations were sufficiently precise, we could distinguish between these two models, especially when the exponential growth phase was well under way. In real scientific and engineering work, however, uncertainty always shrouded the data. The more uncertainty, the longer we'll have to wait before we can distinguish between two models, or between the effects of structure and input.

Consider, for instance, the economic concept of "structural unemployment." A "perfect" theory of supply and demand predicts unemployment that behaves in a certain way, just as a perfect theory of exponential decay predicts a certain curve for disappearance of an aggregate. When actual unemployment doesn't respond to input (demand) the way a perfect theory predicts, the structuralists attribute the discrepancy to structural unemployment—which includes such factors as skill levels, education, age, sex, race, and mobility.

In other words, the structuralists say that *people,* not mass points are out there looking for work. The proponents of the demand theory, however, tend to pooh-pooh these structural factors and attribute the lack of output to lack of *input* (demand). It's not a matter of who is "right," and especially not a matter of who is the crybaby liberal and who the redneck conservative. It's a matter of whose model, on more detailed examination over a longer time period, predicts better. In economics, the split is likely to continue for a while.

In psychotherapy, the same division exists in the distinction between "organic" and "functional" problems. At the organic extreme, the proponents would say that all "mental illness" can be traced to structural malformations in the brain—organic brain damage, senility, chemical imbalance, and the like. At the other extreme, proponents would argue that only a few exceptional cases, such as a large hole in the head, are explained by organic problems. In their view, "mental illness" results from a perfectly normal brain being placed in an abnormal or unhealthy environment. Naturally, the "treatments" prescribed by either economists or psychotherapists will depend heavily on their position on the "structure versus input" spectrum.

A good generalist in any of these fields would have long since recognized that such endless debates are symptomatic of lack of content. Even without knowing the Principle of Indeterminability, a generalist ought to understand that "structure" and "input" must somehow be equivalent for the debate to have lasted so long. With so much heat, any *real* fuel would have been exhausted decades ago.

The resolution of the conflict lies in the understanding that structure is *defined in terms of input*—if you *know* what the input and output are. Conversely, input can be *defined in terms of structure*—if you know what the structure and the output are. In fact, though it's not generally done, output can be defined in terms of structure and input—if you know what both of them are. But, of course, *nobody* can *know* what any of these things are, in the sense of absolute certainty, though most people imagine they are certain about matters within their special territory. Only fools are certain—we think.

Can a Linear System Be Stable?

> All the rivers run into the sea,
> yet the sea is not full.

> Ecclesiastes 1:7

Each of the simple linear systems we've looked at so far has eventually either decayed to nothing or grown exponentially. In studying such systems, then, are we making any progress in our attempt to understand *stability?* Or must linear systems like these always decay or explode? If the answer is yes, then linear models aren't going to be very suitable for the kinds of systems we want to study.

The question of stability of linear systems provides another opportunity for *qualitative* reasoning. We already know that the limiting behavior of any *single* linear aggregate will be exponential. Whether it increases or decreases exponentially will depend on the value of p in the equation:

$$N_{t+1} = pN_t$$

If p is greater than one, the aggregate will grow without bound. If less, the aggregate will decay exponentially. Only in the trivial case of $p = 1$ will the aggregate be stable. Indeed, it will be *immobile*—not a terribly interesting model.

What about differentiated aggregates? In the two-component system of Figure 8.1, with structure matrix

$$\begin{matrix} .5 & 0 \\ 0 & .9 \end{matrix}$$

we saw that the limiting behavior would seem exponential. The same would be true for any matrix of the same general form, with both diagonal elements less than one.

If both diagonal elements are greater than one, the final behavior would also seem exponential, as we see in Figure 8.2, with the matrix:

$$\begin{matrix} 1.1 & 0 \\ 0 & 1.2 \end{matrix}$$

In this case, the smaller component doesn't disappear, but the faster growing one will quickly become so much larger that the other will *effectively* disappear.

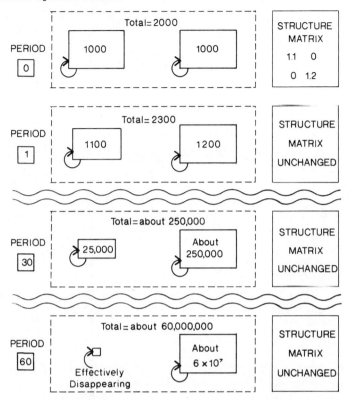

Figure 8.2. When a pair of aggregates each grows exponentially, the faster growing one eventually comes to be so much greater that the other effectively disappears.

If one of the diagonal elements is greater than one and the other is less than one, one will grow exponentially and the other will decay to zero. This can't be considered stable either. Does this answer our question? It almost does for structure matrices with only diagonal elements, but there is one special case we haven't considered. An individual exponential *can* remain steady if the multiplier for each generation is exactly one. If one of the diagonal elements is one and the other is less than one, we would eventually get steady behavior when the decaying component disappeared. In the same way, the matrix

$$
\begin{matrix}
1.0 & 0 \\
0 & 1.0
\end{matrix}
$$

would produce steady behavior in a two-component aggregate. Each component, indeed, would remain *constant,* not much more interesting than the one-component case.

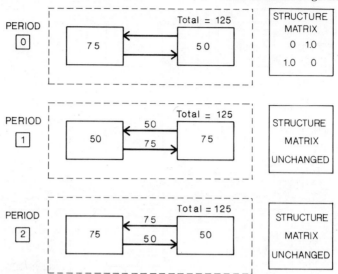

Figure 8.3. When the structure matrix is a "permutation matrix," the system of two or more components will display oscillatory behavior, as in this model of "the changing of the guard."

But with two components, there is another possibility. In Figure 8.3, we see the behavior of a two-component system with matrix

$$\begin{matrix} 0 & 1.0 \\ 1.0 & 0 \end{matrix}$$

applied to an initial vector of (75, 50). This type of matrix, with zeros everywhere except for *a single one in each row and in each column* is called a *permutation* matrix. The reason for the name is clear from the succession of states it produces:

$$(50, 75) \quad (75, 50) \quad (50, 75) \quad (75, 50) \quad \ldots$$

It's not much more interesting than immobile behavior, but it is making progress. Such perpetual oscillation could model a system in which the two aggregates changed roles at fixed intervals. The changing of the guard at Buckingham Palace would be one example, with one aggregate counting men on guard and one counting men in the barracks.

Any permutation matrix will create perpetual oscillations when applied to a state vector, a useful fact when we want to build a model that oscillates in some pattern. But again, the application of simple oscillations is pretty limited. Are there yet *other* kinds of stable behavior in discriminated aggregate systems?

Once again, we can attempt a qualitative approach. For the behavior to be stable in a linear system, the *overall* population must be conserved. That is, it must be equivalent to a renaming system. Looking at the digraph of Figure 8.4, we may ask ourselves: "Under what conditions will the total number of aggregate members be conserved?" By writing out various conservation laws that must hold, it's easy to work out what conditions must prevail.

Consider box number 1. There are three possible ways a member of box 1 can be classified, indicated by the three arrows, A, B, and C. Since the system is to be conservative, each member of box 1 must go somewhere, and none can reproduce. Therefore, since A, B, and C give the percentages going in each direction, their total must be exactly 100 percent. Thus

$$A + B + C = 1.0000000 \ldots$$

is one condition that must hold.

Applying the Principle of Indifference, we see that the same condition must hold for each of the other boxes, and that no other condition is needed to ensure overall balance.

If A, B, and C are to be *percentages* going to each of the three categories, each must lie between 0.0 and 1.0. Other than that, we may

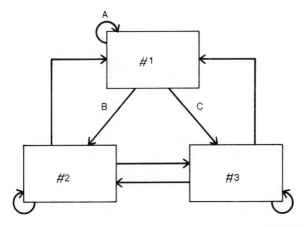

Figure 8.4. To answer the general question, "Can there be stable behavior?" the total number of individuals leaving box #1 (A + B + C) must be exactly 100% of the individuals in the box. The same condition must hold for each other box in the diagram.

choose any combination that sums to 1.0, such as (1.0, 0, 0), (.5, 0, .5), (.01, .001, .989), or (.7, .1, .2). Suppose we make some choices for A, B, and C, and also for the (A, B, C) of each of the other two boxes, percentages which sum to one, shown in Figure 8.5. There we see the digraph labeled, and the structure matrix corresponding to those labels written underneath.

We notice immediately that the condition

$$A + B + C = 1.0000 \ldots$$

is precisely the same as requiring *each column* of the structure matrix to sum to 1.000. . . . Why are they identical? Because column one of the structure matrix gives all the fractions of subaggregate 1 going *to* some subaggregate *from* 1. That is, A of Figure 8.4 was *to* 1 *from* 1, or M_{11}; B was *to* 2 *from* 1, or M_{21}; and C was *to* 3 *from* 1, or M_{31}. The same is true for the other two columns, so the conditions on the arrows are the same as the conditions on the matrix.

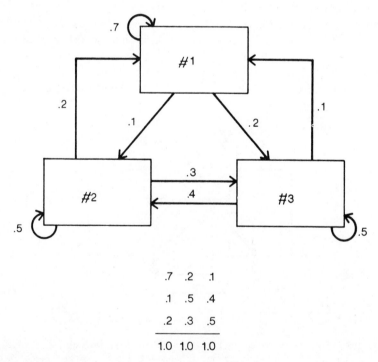

$$
\begin{array}{ccc}
.7 & .2 & .1 \\
.1 & .5 & .4 \\
.2 & .3 & .5 \\
\hline
1.0 & 1.0 & 1.0
\end{array}
$$

Figure 8.5. A stable linear structure with three elements is described by this structure matrix, each of whose columns sums to 1.00000 . . .

Because we have constructed it that way, we know that M will model a renaming system, even before we see the results worked out. Qualitative reasoning thus answers our question. There *can* be stable behavior with a linear structure. That structure need not be simple *immobility* of the subaggregates; nor need it be a simple *oscillation*, as we got with a permutation matrix for M. What, then, does this behavior look like?

We could now turn to our computer, giving it the task of projecting the linear structure forward in time by operating on some initial vector, v, using the program step

$$2. \quad v = M \times v$$

But before rushing to the computer, it's always a good idea to see if more can't be squeezed from our intuition, or from some qualitative reasoning. We know that the total of the three subaggregates will be conserved, so we know that the exact behavior will be different for different initial aggregates. If their total numbers of members differ, they can't show *identical* behaviors, since each will conserve a different total.

Suppose we choose two different initial vectors that have the *same* total, such as $(1000, 0, 0)$ and $(0, 500, 500)$. We know that each will have 1000 members at all times, but do we know anything about how those 1000 members will be distributed among the subaggregates? In order to make this observation simpler, we did choose the elements of M in a somewhat special way. When we look at M,

.7	.2	.1
.1	.5	.4
.2	.3	.5

we notice that not only do the *columns* each sum to one, as required, but also the *rows* each sum to one, by "accident." What would a structure like this mean?

Well, what do the rows mean? Row 1 is all the ways a member can get *to* subaggregate 1, *from* 1, 2, or 3. Row 2 is all the ways to get *to* 2, and row 3 is all the ways to get *to* 3. Each of these rows adds up to the same thing, namely one, or 100%. Each of the ways *out* of each subaggregate also adds to the same thing, one, or 100%. It's not very mathematical, but our intuition tells us we might be able to guess something about these subaggregates. In some way, they're all the same. None seems more heavily committed to input or output than the others. Could it be that in the long run, they'll each have the *same* number of members?

Mathematicians and scientists guess all the time, though they rarely admit it except when their guesses turn out right. Even then, some credit their discoveries to "rational thought." Guessing is a good instinct to develop, as long as we develop the twin instinct of *checking* our guesses. In this case, nothing could be easier. What happens, say, if the initial vector has all three subaggregates equal, as in

$$(100, 100, 100)?$$

First checking v_1, we get

$$.7 \times 100 + .2 \times 100 + .1 \times 100 = 70 + 20 + 10 = 100$$

Oh, how stupid of us! If all three subaggregates are the same, then each element in the row is multiplied by 100, so v_2, for instance, is

$$100 \times (.1 + .5 + .4) = 100 \times 1 = 100$$

In other words, since each row adds up to one, and each element is exactly 100, each element *stays* at 100 after applying M to v. Thus, our guess is confirmed. An initial vector of $(100, 100, 100)$ will be *unchanged* by the program represented by this matrix.

A harder question is this. What happens if the initial matrix isn't $(100, 100, 100)$, but something *close*, like $(101, 100, 99)$? We know the total of 300 will be preserved. Can we guess again, this time that the program will bring the vector back to $(100, 100, 100)$?

It's a bit more complex to verify this guess with a simple argument, but it's true. In fact, no matter *what* the initial vector is, if it sums to 300, the final vector will be arbitrarily close to $(100, 100, 100)$. If, when it reaches $(100, 100, 100)$, we move a member or two from one subaggregate to another, it will go *back* to $(100, 100, 100)$ in short order. Thus the system is stable not just in the sense of remaining unchanged, but of recovering itself from changes introduced from the outside.

Why Linear Systems Are So Popular

Vectors find applications in many parts of mathematics. . . . Topics in which vectors are used extensively include mechanics, dynamics, linear differential equations, statistics, differential geometry, mathematical economics, function spaces, algebraic number theory, systems of linear equations, and theory of games. (Thrall and Tornheim 1957:1)

This quotation is taken from the first page of a text that was used to educate a generation of would-be scientists in the pleasures of linear systems.

It comprises essentially all of the justification for studying over 300 pages of dense axiomatic development. Other textbooks in use then and now generally use even less space bothering to inform the student *why* this subject is being studied so strenuously for one or two semesters. Although the list of "applications" may make the mathematician's mouth water, most 19-year old sociology majors—even if they should find themselves in such a course—would not know the meaning of at least half of the "applications" in the list.

It's no wonder, really, that biological and social scientists are "turned off" by mathematics, and think it to be a difficult subject. From the first grade onward, their earnest and sincere question, "What good is this stuff we're learning?" is turned away with scorn, like a doctor scolding a patient who asks what the bitter syrup will do for the pain. "It's *good* for you." Indeed, we know it's good for us *because* it's so bitter. Actually, most mathematics teachers don't really know why they're teaching what they're teaching. So perhaps we can cast a little light on the subject.

In Figure 8.6, we see the initial two periods of the system with structure matrix

$$\begin{matrix} .7 & .2 & .1 \\ .1 & .5 & .4 \\ .2 & .3 & .5 \end{matrix}$$

and the initial state vector, (1000, 0, 0). The members are shuffling around, and the total of 1000 members is conserved.

In Figure 8.7, we see the three subaggregates plotted over time, showing that after about 10 periods, they settle down to the constant value (333.3 . . . , 333.3 . . . , 333.3 . . .). To confirm our prediction, Figure 8.8 shows the same system started at (0, 0, 1000). This is about as different a state as we can get from (1000, 0, 0), but it, too, leads to (333.3 . . . , 333.3 . . . , 333.3 . . .) after a few periods. This state is called the *equifinal* state for all initial states whose elements sum to 1000. The equifinal states for other states are similar, being composed of three equal elements, but larger or smaller depending on the initial state.

As it happens, the equifinal state can be deduced in advance for *any* linear system from its structure matrix. Indeed, one of the reasons the linear system is a favored model is that so much can be deduced qualitatively, or with simple algebraic work, about linear systems. We chose the example of Figure 8.6 because it allowed us to reason more easily about its equifinal states, but actually we didn't save much trouble. Consider a more general matrix, in which the rows do *not* each sum to 1.0000.

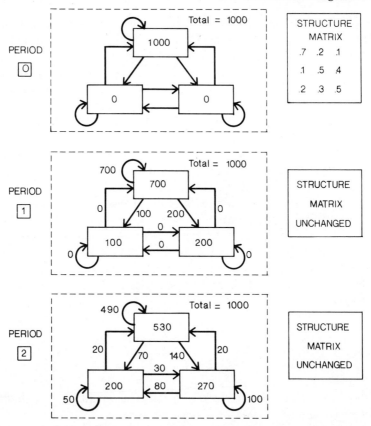

Figure 8.6. These numerical results confirm that the system of Figure 8.5 produces a constant overall population, as predicted by the way the matrix was constructed.

Take, for instance,

$$.4 \quad .4 \quad .5$$
$$.2 \quad .4 \quad .5$$
$$.4 \quad .2 \quad 0$$

Since the *columns* sum to one, we know there will be an equifinal state. Since, in the equifinal state, none of the subaggregates change, we can write three equations for (v_1, v_2, and v_3). For instance, when the equifinal state is reached,

$$v_1 = .4v_1 + .4v_2 + .5v_3$$

This merely says that the three contributions to the new v_1 will simply add up to the same as the old v_1—which defines the equifinal state. By

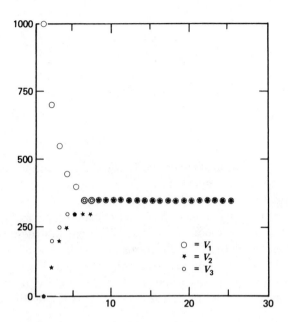

Figure 8.7. With initial population (1000, 0, 0), the system comes to an equifinal state at (333, 333, 333).

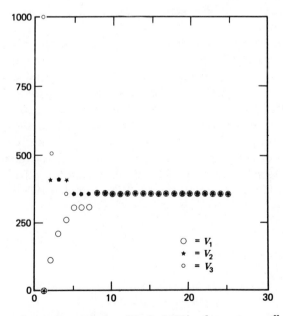

Figure 8.8. With initial population (0, 0, 1000), the system still comes to an equifinal state at (333, 333, 333).

155

writing down all three of these equations, and then solving by simple algebra, we find that in the equifinal state, the subaggregates are in the proportion

$$(25, 20, 14)$$

Testing this on v_1, we get

$$v_1 = .4 \times 25 + .4 \times 20 + .5 \times 14 = 25$$

Knowing how linear systems work, we can not only *analyze* linear models, but we can *design* them to fit our needs. The job is so well understood and straightforward, we can even establish a completely automatic computer program that takes a set of data points as input and produces a "fitted" linear model as output. When we add these conveniences to the convenience of being able to make a multitude of qualitative deductions and to the convenience of having a standard computer program that can project any linear system forward in time, we comprehend the enormous favor lavished on linear systems by systems thinkers all over the world.

Because they are so convenient, we often forget *why* we chose linear models. More than one systems thinker has been lulled into dreaming that the world about us is somehow "really linear." *We* know, however, that modeling is simply a game we're playing, though often for high stakes. Linear systems are *one* way to play the game, a way whose favorable properties make it the first choice to try on a new problem. But we also know that playing the linear game in the wrong situation can create more problems than it solves.

We know, for example, that linear systems have the wonderful, but very special, property that allows the structure part, *M*, to be cleanly separated from the behavior part, *v*. Certain nonlinear systems also have this convenient property of clean separation. But not all systems need be, or are, so neatly separable. In some—in most, we sometimes think—the separation doesn't exist at all. Therefore, we mustn't take linear systems for granted, but must ask ourselves, "Why is it that *this* system happens to be modeled quite nicely by a linear system?"

The Structure-Regulation Law

> Population, when unchecked, increases in a geometric ratio.
>
> Thomas Robert Malthus
> *Essay on Population*

Now that we've raised the question, why *is* it that in some systems we *can* separate a structure part from a behavior part? First of all, such a separation is possible only if the "structure part" remains relatively stable. Otherwise, it wouldn't be "structure"—that which stands—at all, but part of the behavior.

But system parts don't just remain stable without reason. It is the incessant process of *regulation* that keeps them stable. But conversely, it is their stability—their structure—that makes regulation possible in the first place!

The casual observer in the workaday world may easily overlook this reciprocal relation between structure and behavior. To keep it more toward the front of our minds, we're going to elevate it to the prestigious position of a law. *The Structure-Regulation Law* says the following:

> **Stability is made possible by the process of regulation; regulation is made possible by the existence of stability.**

The structures around us often *seem* to remain stable "all by themselves." Barrington Moore, in the quotation which began this book, attacked this fallacy in the social sciences—a position with which most of us can readily agree. But the Structure-Regulation Law applies equally to inanimate structures, such as rocks or walls. Although rocks and walls may not exhibit, to us, such "active" regulation as salamanders or states, they are regulating all the time. If we lean on a wall, it must fall over from this stress, or input, unless its structure can accept the input without collapsing. As any mountain climber knows, a rock warmed by the sun or cooled by a frost will crack or shatter unless prevented by its structure.

Rocks and walls are regulating just as living beings or social systems are regulating. Those who believe otherwise say that a large part of the world as we happen to see it is an *inevitable* consequence of structure, as in a structure matrix such as

$$
\begin{array}{ccc}
.7 & .2 & .1 \\
.1 & .5 & .4 \\
.2 & .3 & .5
\end{array}
$$

But the structural stability represented by such forms is a kind of deception, for it conceals the true problem. How does it happen, we should ask such believers, that the numbers in that matrix hold onto those values so precisely? Why are *they* stable? Where did those numbers come from anyway?

Engineers who work with control systems define a quantity called "sensitivity," which, roughly, measures how much the behavior of a system varies when some part of the system is varied. The part may be part of the state vector, in which case the sensitivity is a measure of what is called *Liapunov stability* (Tomivič and Vukobratovič 1970). This is the sort of stability we've seen in Figures 8.7 and 8.8. Even a large displacement of the state vector can be brought back to the equilibrium position by the action of the system's linear program. This system is not "sensitive" at all.

But the same cannot be said for the response of the system to a slight change in one of the elements of the *matrix*. To be stable in the face of this kind of change is what the engineers call "structural stability." The Law of Effect (Weinberg 1975) expresses the *hope* that "small changes in structure will lead to small changes in behavior." If that hope is realized, it will obviously be easier to study systems, for "similar" systems will most often exhibit "similar" behavior. But the Law of Effect is not really a law as much as it is a *hope,* and we must never forget it.

Thorstein Veblen once observed, "How slender an initial difference may come to be decisive of the outcome in case circumstances give this initial difference a cumulative effect." But if this "initial difference" is in the systems *program,* it will presumably be acting *again and again* on the system's state. Thus it will be very likely to produce a cumulative effect.

To see what Veblen meant, let's consider the system of Figures 8.7 and 8.8 with a small change in its *matrix,* rather than in its state vector. The matrix, of course, was that part of the system we had assumed "fixed," so it wouldn't be fair to make a *large* change. To see if this system exhibits *structural* stability, then, let's just change one of its elements, M_{11}. And let's just change that element by the *tiniest* amount, from .70 to, say, .69. Certainly any system worthy of the term "structurally stable" ought to be able to remain relatively undisturbed by that small a change.

Presenting old and new side by side to facilitate comparison, the matrices look like this:

	OLD			NEW	
.7	.2	.1	.69	.2	.1
.1	.5	.4	.1	.5	.4
.2	.3	.5	.2	.3	.5

What happens when the new matrix is applied to the vector (1000, 0, 0)? The answer can be seen in Figure 8.9. For a while, indeed, this behavior is very similar to that of Figure 8.7, but around period 9, first one age group and then the others begin to fall. If the curves were extended

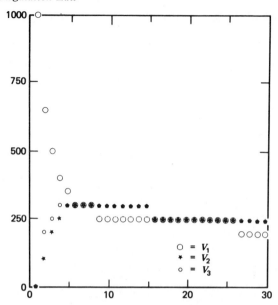

Figure 8.9. By changing one column sum from 1.00 to .99, we make the system lose its stability. It starts out like the previous system, and behaves similarly for a while. Eventually it begins to lose population. The equifinal state is (0, 0, 0), not (333, 333, 333).

further in time, we would see the completion of a slow exponential decay, not at all the behavior of the *old* matrix. Why does the system decay exponentially? Because the initial difference of .01 in M_{11} has a "cumulative effect."

If we had *added* the .01 to M_{11}, the results would have been somewhat more spectacular—the exponential *growth* that we see just starting in Figure 8.10. Exponential growth is spectacular partly because we don't often see it exhibited by real systems. But if we don't, then most "linear" systems must have something hidden in the background that stabilizes the matrix itself.

For purposes of mathematical modeling, we may *pretend* that these hidden "somethings" aren't there, just so long as we don't forget the Structure-Regulation Law. But from an engineering point of view, there's no particular reason to distinguish the numbers in the matrix from the numbers in the vector. *Anything* that changes any part of the system is a potentially damaging "input." The matrix part in a linear system seems *particularly* vulnerable to such change.

When engineers or social planners build their linear idealizations into physical realities—a communication network, a chemical processing plant,

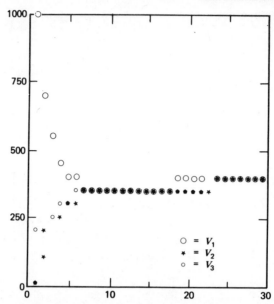

Figure 8.10. By changing one column sum from 1.00 to 1.01, we make the system lose its stability and turn into an exponentially increasing population. Again, it starts out like the previous systems, but eventually it starts to grow. There is no equifinal state, for all three components will become infinite.

a traffic control system, or a health-care delivery system—they must plan that the components will never be quite so fixed as their linear model had to assume. Resistors change in value or "drift" over time. Valves get sluggish or don't quite close. Timing relays sometimes skip a step, or "bounce" and introduce an extra step. Patients take all their pills at once, or use a suppository as a pill.

Quite often, the difference between a successful and an unsuccessful engineered system is precisely in the successful system's low sensitivity to real-world "accidents" or "decay." Usually, the designer has to sacrifice some "efficiency" to get such structural stability, and the system may not seem to do such a good job when working. But if it keeps working when alternative systems fail, it may be the only *possible* system for us to consider applying in a critical situation.

Probably the greatest single cause of computer system "disasters" lies in the failure of their designers to consider structural stability. Simple calculations are made to show the "feasibility" of the new system, then these calculations are refined to show how every last drop of "efficiency" will be squeezed out. Unfortunately, structural stability is never considered in these "refined" calculations, so it may be squeezed out along with the efficiency drops. Things run so efficiently that the designers win awards. Then comes the first "structural" failure, usually blamed on the operators, rather than the designers.

But it is the designers who—in the rush for glory—have forgotten to consider the Structure-Regulation Law. The prudent designer—the true systems thinker—will interrupt thoughts of glorious "efficiency" long enough to ask, "What regulates the structure matrix?" Even more generally, the designer will put the Structure-Regulation Law into the form of a question paraphrasing the old Roman dictum:

"Who regulates the regulator?"

In this form, the law has great heuristic value and will lead our thoughts out from behind the curtain with which they are so often veiled. Though a certain birth rate may stabilize a population, what stabilizes the birth rate?

People who believe that structure matrices stay constant "because they just do" probably also believe that rocks do not regulate and that the "solid" world around them does not move. General systems thinking may not be terribly empirical, but it does require a certain minimum of observational awareness. In fact, potential systems thinkers might test themselves by asking a few "simple" questions, such as:

"Do plants move?"
"Do rocks move?"

Those who fail this part of the test might try spending some time in the woods with a copy of *Walden*—for quiet contemplation of the environment.

The physical structures in our environment *do* move in response to forces acting on them, though the magnitude and time scale of their movement may be out of range of our casual observation. Even the pressure of a hand on a wall moves that wall, as can easily be detected with an interferometer. When the wall moves, its state changes, but within a range acceptable to us. When we remove our hand, the wall may move back, either all or part of the way. Because the movement is so small, and because many leanings will not accumulate much distortion in the wall, we simply imagine that the wall is "solid" and stationary, that it does not regulate.

Perhaps the most striking evidence of regulation by "solid" objects comes when that regulation fails. A wall falling is most spectacular, but a more impressive demonstration may be seen, for instance, on the steps of ancient temples. For the steps, the regulatory problems posed by a single humble pair of human feet or knees is not great, and the builders of the temple probably thought that their granite was immortal. But, over thousands of years, and millions of feet each year, the steps begin to show that they cannot quite—on this time scale—perform the regulatory job for which they were built. Crumb by crumb, step by step, they turn to dust and are carried off by the wind.

QUESTIONS FOR FURTHER RESEARCH

1. *Applied Mathematics*
   Study the behavior of a two-component aggregate with initial state (100, 100) and each of these three structure matrices:

$$\begin{array}{cc} 0 & 1 \\ 1 & 0 \end{array} \qquad \begin{array}{cc} 0 & 1.01 \\ 1 & 0 \end{array} \qquad \begin{array}{cc} .01 & 1 \\ 1 & 0 \end{array}$$

Discuss the nature of the stability of such systems, and how it relates to their value as models.

2. *Economics-Sociology*
   An economic theory relating land, labor, and money in a simple market model fails to account for the "resistance" that social structure imposes on the "free movement" of such things as prices, wages, and rents—in the same way that the structure of a bridge resists the simple acceleration of gravity. Although economics considers itself "the queen of the social sciences," social structure seems to be an anathema to economists. Discuss the following proposition from the point of view of economics:

Social structure is that which makes pure economic theories fail.

Then discuss the same proposition from the point of view of the sociologist.

3. *Demography*
   Lotka (1924) demonstrated that any population governed by a linear model would exhibit a certain kind of stability, even if it grew exponentially. The stability was not of population *size*, but of age distribution in the population vector. In other words, the *relative* numbers in each age bracket would stay the same, even if the overall population increased or decreased. Therefore, if one observes a stable age distribution, one can reason backward to a linear model approximating that behavior. From the matrix, one can select values of characteristics of the population, such

as birth rate, death rate, and growth rate. Of course, many linear models fit, but usually some idea of one or more of these rates is available, from which others can be inferred. To do this kind of work, demographers use tables of stable populations corresponding to different rates of increase and levels of mortality. Try to use these tables to infer some of the characteristics of the United States population between 1950 and 1970, using the data from Research Question 5 of Chapter 7. What might be wrong with these inferences? (Reference: Coale and Demeny 1966)

4. *Anthropological Demography*

When the population of a country is modeled by a linear model, an influx of immigrants in some time interval can be represented by adding appropriate numbers to the population vector. If the immigrants are mostly old people, their effect on the population will be transient, because they will do little reproducing. On the other hand, if they are young, they may add disproportionately to the births in a few years. In either case, the effect of the immigration will be "melted" into the population after a generation or two, just as the "melting pot" model suggests.

But suppose these immigrants affect the culture itself. Old people may bring new ideas on birth control, positive or negative, which can be transmitted to the nonimmigrant population. Young people might convey new attitudes about early marriage or family size. Such changes in culture will be reflected in the matrix of the linear model, not in the vector, at first. Compile a list of as many such cultural factors as you can find by which immigrants could affect the country's population behavior. For each one, roughly quantify the effect and explain how its change in the matrix could be modeled. Compare the long and short term effects of the two kinds of changes that an immigrant group can foster.

5. *More Anthropological Demography*

Discuss the question of whether or not *emigrants* can affect the country's matrix. If they can, repeat the previous question for emigration.

9

The Search
for Regulation

No one will quarrel with the assertion that social existence is controlled existence, for we all accept a certain basic assumption about human nature—namely,
that without some constraint of individual leanings the coordination of action
and regularity of conduct which turn a human aggregation into a society could
not materialize. (Nadel 1968:401)

A pure aggregate strategy may be all right for auks, aubergines, or au-
riferous rock, but human societies from the australopithecines to the
Austro-Hungarian Empire have some "coordination of action and regularity of conduct." No person who has felt "the constraint of individual
leanings" can lack curiosity about the deep reasons and strategies behind
the regulation necessary for survival—regulation above the level of a mere
summation of individual activities. In this chapter, we begin to move
away from description of aggregate behavior toward understanding some
of the structured strategies that have been built upon them.

The Problem of Multidimensional Regulation

The study of equilibria will always be important in the treatment of systems of
high complexity, for the equilibria, in their various forms, are those states, or
sets of states, in which the system's behavior no longer depends to a major
degree on the time. By effectively losing a variable, the functional relation becomes simpler; and the change may reduce the impossibly complex to the
manageable. (Ashby 1964:95)

Concealed by our impression of a relatively stable, structured world is
the unceasing regulatory activity of aggregates. Aggregate survival is the
most *elementary* form of regulation, both in the sense of the simplicity

164

with which it may be understood and modeled and in the sense that it is the foundation on which other regulatory mechanisms are built. As Ashby indicates, where there is such a great complexity in the environment, it is advantageous to be able to "lose a variable," so that the problem of regulation can be made tractable. Relying on aggregates, then, is one of the simplest ways of losing variables.

But why should "losing a variable" be such a helpful action for a system to take? Engineers are well aware, or should be, of the simplifying effect of removing one of the constraints from a problem. Engineers are annoyed by the introduction of such "nonengineering" concepts as beauty, not so much because they are against beauty but because people asking for beauty usually think that beauty is something that can be "painted" onto the otherwise strictly engineering solution. They do not realize that for a truly beautiful bridge, beauty must have been an integral part of the design—and the cost estimates—from the beginning.

To see how the addition of dimensions affects the design problem, let us consider a *very* hypothetical design problem given to a lighting engineer. The present streetlighting system in our neighborhood has proved to be unsatisfactory because the young girls living there have recently acquired BB guns and like to use the lights as targets. The village cannot replace bulbs fast enough to keep up with the girls, and because the constitution guarantees the right of the populace to bear arms, it has proved impossible to legislate a prohibition.

The village elders come to the lighting engineer for help, and she begins by listing the various forces in the environment that will affect the lifetime of a streetlight. For simplicity of our argument, suppose the list contains only two items:

1. The BB guns.
2. Heat generated by the bulbs themselves.

The engineer then finds out that the effects of each of these two disturbances depend on the thickness of the individual bulb, T. A thin bulb dissipates heat quickly and is thus better able to survive the destructive effect of its own internal heat. A thick bulb, on the other hand, has a better chance of surviving a hit from a BB. Moreover, there is a limit to the thinness of the bulb, below which it will shatter spontaneously, and to the thickness, above which it will transmit no light.

The problem facing the engineer is to design a system of lighting that will supply a certain minimum amount of light over a certain minimum time—long enough, at least, so that replacements can be made on a reasonable schedule. As in all engineering problems, however, there is one more constraint—*cost*. If the village elders are willing to put up enough cash, the engineer can simply employ a pure aggregate strategy and design a very large light stanchion containing thousands of bulbs. Given that the girls cannot afford infinite numbers of BBs, nor infinite time to be shooting at streetlights, such a pure aggregate solution would guarantee a minimum of light for some minimum time.

With a cost factor, however, the lighting engineer is constrained to some maximum number of bulbs per stanchion. This we can call N_o. Just to keep our problem within the bounds necessary for exposition, let us assume that there are no better bulbs available and that the engineer has no light-transmitting material available other than the bulbs themselves. Thus she can cluster the bulbs in any way she likes, but she cannot introduce other materials into her solution.

Given that she has a maximum of N_o bulbs to work with, the engineer might decide to adopt some mixed strategy, using some bulbs of one thickness, some of another, perhaps a few of a third. Before considering such mixed strategies, however, she would probably examine the expected behavior of the two pure cases—all bulbs of maximum thickness or all bulbs of minimum thickness. She might calculate, or obtain from the manufacturer, a chart something like Figure 9.1, showing the rela-

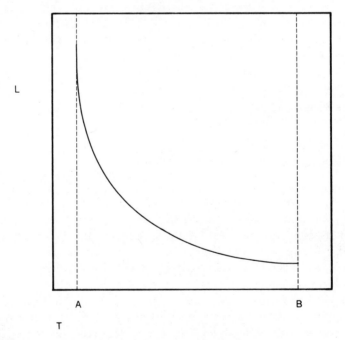

Figure 9.1. Lifetime (L) of a hypothetical lightbulb as a function of thickness (T), assuming the only cause of failure is burnout of the bulb.

tionship between thickness of the bulb and expected lifetime resulting from internal heating. She might then do some calculations or experiments of her own to obtain Figure 9.2, showing how the lifetime is affected by thickness under a fusillade of flying BBs. These charts were obtained from explicit computer models, but their details are of no concern. Notice only that lifetime *decreases* with *thickness* because of heating and *decreases* with *thinness* because of BBs.

These two charts verify our intuitive feeling that if there were no girls to contend with, the engineer should choose the minimum thickness, A. If, on the other hand, internal heat were no problem, B would be the thickness of choice. To determine what would happen in a *combined* environment, however, she would have to merge the two causes of death into such a chart shown in Figure 9.3.

Using this graph, she could choose the optimal thickness for the bulbs (the point designated as U). This maximum falls between the two extremes, A and B, because of the nature of the problem posed to the system by this combined environment—and because of the specific parameters of the problem. But more important than the *location* of U is its

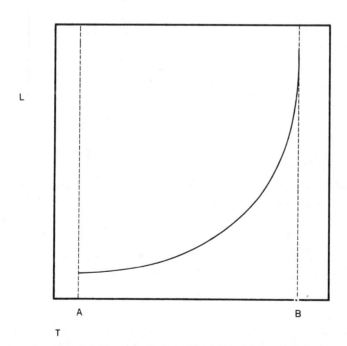

Figure 9.2. Lifetime (L) of a hypothetical lightbulb as a function of thickness (T), assuming the only cause of failure is shattering from BBs.

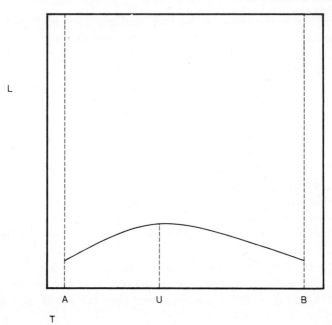

Figure 9.3. Lifetime (L) of a hypothetical lightbulb as a function of thickness (T), taking into account the two sources of failure, burnout and shattering from BBs.

size—for our model the value at U was 400 hours, compared with about 80,000 hours for the values at T_{max} or T_{min}.

Why does the combined environment result in an optimum survival time two orders of magnitude smaller than the survival time in either of the two simple environments that compose it? The answer lies in the conflict forced on the system by the multiplicity of the environment. The graphs in Figures 9.1 and 9.2 show that any movement in the direction of improvement in one environment is rapidly met with a compensating loss in performance if faced with the other. Since the two must be dealt with simultaneously, the best the system can do is rather poorly in both at once. This, then, is the problem of multidimensional regulation:

> **When a system has to deal simultaneously with two threats, protection against one will increase vulnerability to the other.**

Indeed, if this were not true, then we wouldn't have *two* threats, but only one disguised as two.

Naturally, the force behind the problem of multidimensional regulation will depend on both the exact *nature* of the threats and their *number*. We shall consider these aspects in turn.

Separation of Variables

> All work, and no play,
> Makes Jack a dull boy.
> All play, and no work,
> Makes Jack a mere toy.
>
> Primary Language Charts
> American Book Co., 1892

When faced with the problem of multidimensional regulation, we naturally wonder whether we could somehow arrange a system that will get the *best* of each environment, rather than close to the worst. In an attempt to do just this in the BB problem, our engineer constructed a system in the following manner: the bulbs were arranged in two *layers*, one of which completely covered the other, as shown in Figure 9.4. Since the bulbs were transparent, the covering layer did not block the light from the other layer, but was sufficiently thick to block the BBs. This outer layer, then, could solve *one* of the regulatory problems facing the system, leaving just one *other* problem—the heat problem—for the inner layer.

Let us see in more detail how this solution works. Since one outer bulb is needed to shield each inner bulb, we can afford only half as many bulbs ($N_o/2$) in each layer. In a reasonably short time (about 300 hours) the outer layer will be all burned out, because of the effects of heating. The *system*, on the other hand, will still be performing very well—very well indeed, because no more than one or two of the inner bulbs will be extinguished.

Calculation of the lifetime of the protected inner bulbs shows that the expected lifetime (until, say, only $N_o/3$ are left lit) is about 30,000 hours, even though there were only $N_o/2$ bulbs to begin with! By sacrificing *half* its bulbs to form a protective layer, this system has extended its useful life by several orders of magnitude. Viewed another way, a system of identical bulbs would have to cost 100 times as much to provide the same lifetime as this separated system.

The strategy of *separating the variables* in the environment and providing different mechanisms for dealing with each is almost universal among systems that are successful survivors, as we might well expect from the cost advantages. A business firm requires information about its activities for its stockholders, but it also requires that its competitors do not know too much about what is going on. This conflict, a striking parallel to our lightbulb example, sometimes tempts a firm to keep two sets of

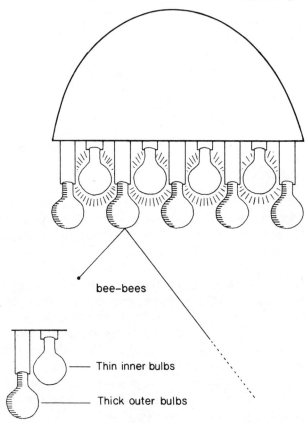

bee–bees

Thin inner bulbs

Thick outer bulbs

Figure 9.4. A hypothetical two-layered streetlight in which the thick lower bulbs burn out quickly, but protect the thin upper bulbs from BBs.

books: one to deal with internal needs and one to show to the outside world. Though this strategy may be illegal, it is at least a classical case of separation of environmental variables. In government, it is called "diplomacy."

The components for handling each environmental variation need not be physically separated. Mammalian blood, for instance, is a mixture of several types of cells and other components, most of which are specialized to meet particular threats to which the body is subject. White cells, for instance, are called on to engulf foreign particles entering through a break in the skin. The antibodies in the blood form the most striking example of separation of threatening variables, because the immunological mechanisms manufacture a new type of antibody for *each* foreign protein type that invades the body. Here, the general advantage of separating sources of variation has been elevated to a general mechanism for producing a new type of defense as each new type of threat is encountered. The existence of such a technique lends its possessor a substantial advantage in the competition for survival in a world of multiple threats.

Like the aggregate strategy, the separation strategy is so universal that we may take it for granted or not even notice it. As our Earth rotates about the sun, its tilted axis causes *seasons*, an alternation of environmental problems for any system on Earth. Our houses have both heating and cooling systems; our clothing comes in winter and summer versions; our skin turns light and dark on exposure to the sun. In summer, plants produce seeds that survive the winter. Hibernating animals become inactive during winter; desert animals, in summer.

The rotation of Earth about its axis causes *days*, an alternation of light and dark environments. We have sunglasses and florescent lamps. Hawks feed by day and sleep by night, and owls reverse the strategy.

The surface of the Earth provides three obviously different environments—air, land, and water—and many species operate in two or even three. Each larger environment breaks down into microenvironments—saltwater, freshwater, and estuaries that alternate between the two. Each microenvironment is broken down again, and those species operating in two or more reap the same advantages of separation that accrue to those that separate the larger environments. With such a ubiquitous strategy, those advantages must be enormous.

The First Law of Regulatory Compromise

> To be, or not to be: that is the question:
> Whether 'tis nobler in the mind to suffer
> The slings and arrows of outrageous fortune,
> Or to take arms against a sea of troubles,
> And by opposing end them? To die: to sleep;
> No more; and by a sleep to say we end
> The heart-ache and the thousand natural shocks
> That flesh is heir to . . .
>
> Shakespeare
> *Hamlet* III, I

Just how *much* of an advantage does separation of variables give? We cannot expect to give a *precise* answer to such a question, but we can make a start by asking another. *How many threats* does a system face?

Hamlet thought there were a "thousand natural shocks that flesh is heir to," but for the human being this is certainly a low estimate. We need only consider the myriad insects, bacteria and viruses that might plague us if it were not for our regulatory systems which make most of them invisible. But, in any case, the number would depend on how we decide to count a "threat," and we know how such a count might vary from observer to observer.

The really important thing is to realize that for systems of any complexity, "thousands" is a gross underestimate. Even for such an artificial system as a moon rocket, the list of possible malfunctions runs into thousands, and then some malfunction inevitably occurs that is not on the list.

In our lightbulb example, separating the two regulatory problems increased the performance of the system a hundredfold. Is this a typical figure for such a separation? Again, no answer can be both precise and general at the same time, but a factor of 100 is not atypical for certain engineering problems. On the other hand, we know that in other cases the advantage is not so great. Perhaps one way to guess intelligently about this problem is to recall our experiences with computations. For some kinds of regulation, as we shall see, something like a computation is required. Therefore, we might consider that the Square Law of Computation would give us some feeling for the power of separation of variables in a regulatory system.

If the difficulty of regulation went up as the square of the number of separate problems (n^2) presented by the environment which the system must handle simultaneously, then separation of variables should improve things by a factor of n. This in itself would explain the importance of the separation strategy, but the difficulty of regulation could easily be dependent on some stronger function of n, such as 2^n. In this case, dividing 1000 regulatory problems into two sets of 500 would simplify the problem by a factor of 2^{499}, or about 10^{158} times!

Whatever our crude estimate of the power of separation of variables, there are other factors that would tend to reduce that power. In the first place, the environmental threats are *not all independent*, so that in some cases many threats can be handled adequately, if not perfectly, by a single mechanism. The immunological system can be seen as a single mechanism for handling a wide spectrum of threats. The protective ring behavior the musk ox adopts against wolves will certainly protect the beasts equally well against other predators that happen to be in the neighborhood. Thus we must take into account a certain advantage of generalized mechanisms dealing with *related* threats.

Of course, the most general of the generalized methods of regulation is *aggregation*, on which most other strategies ultimately rest. But, to use the aggregate strategy, the system must not be too differentiated, according to the Second Law of Aggregate Survival. In other words, the requirement for structured regulation is just *complementary* to the requirement for redundancy of parts, and the actual mechanisms employed by a system must strike a balance between the two.

An example of such a balance often found in physical systems is that between compactness and dispersion. A herd of animals can achieve the regulatory advantages of coordinated behavior only by remaining rather close together. In doing so, however, they are likely to find themselves all exposed to the same environmental threat, such as a landslide or a disease, which some members could have avoided if they had remained dispersed. As another example, the living cell makes coordinated chemical activities possible within the confined scope of its cell wall, but in doing so must necessarily expose itself to complete destruction if it encounters some very unfavorable local condition.

The compromise between the generalized power of aggregate survival and the specialized power of separate structures for separate problems is so pervasive that we wish to give it a name. Stated concisely, the First Law of Regulatory Compromise is this:

Aggregation gives protection against the unknown;
specialization, against the known;
and the use of each sacrifices some opportunity to use the other.

The Second Law of Regulatory Compromise

> A. What glows in the dark, hangs fifteen feet
> above the ground, and weighs a ton?
> B. A flying saucer?
> A. No, two 1000-pound streetlight bulbs.
>
> After a popular riddle

The advantage of separating variables is great enough to ensure that any system surviving in a multiple-threat environment must have made such separations. Yet we would not expect separation to be a universal strategy unless multiple threats were universal. Do all environments present multiple threats to systems?

In light of the arbitrary system-environment boundary, the answer would at first seem to be "no." To be sure, systems that survive are not quite so arbitrary; but, still, what threatens one system may be a blessing to another. You might be exceedingly unhappy in a room full of mice, but your pet snake would be ecstatic.

In the case of our lightbulb system, what would happen if someone discovered a bulb that thrived on heat, so that its life expectancy got longer as it got thicker? Using such bulbs, our problem would become one-dimensional, and we would just choose the thickest possible bulb to solve both problems at once.

Well, then, as long as we are choosing the thickest possible bulb, why not choose one 40 feet thick? Forty feet of glass would protect the bulb from a direct hit by a howitzer shell and hold in more heat than a lime kiln. The lifetime of such a bulb should be practically infinite. But surely we can imagine many reasons why a bulb 40 feet thick would not be satisfactory. Very little light would penetrate the glass, and we would need too many bulbs to get the minimum lighting. The immense size of the bulb might not permit enough bulbs to be put into a small enough space to provide the light we need. The weight of such a quantity of glass might make it impossible, or unsafe, to suspend the fixture above the street. And perhaps the manufacturing cost for bulbs goes up drastically as the amount of glass increases.

The point is that *for one reason or another*, the thickness of a bulb could not be increased *indefinitely*. Eventually, some *new* disadvantage would become the same order of magnitude as the one we were trying to escape by increasing the thickness. Exactly *which* disadvantage does not matter—there will always be one, because, if there is not, we shall keep increasing thickness until there is. Since no system can be infinitely thick or thin, heavy or light, or simple or complex, the best *surviving* design will always be found at a *compromise point* where at least two threats are more or less balanced. Since only the best survivors will be around to be seen, the systems we see will tend to be found in this type of multi-threat environment, for nothing is infinitely anything. Therefore,

There is a limiting factor to every regulatory strategy.

This is the *Second Law of Regulatory Compromise*.

The Second Law can only tell us that there *will be* a compromise, from which we can infer that every system must deal with a multiply dimensioned environment. The Second Law cannot tell us *which* factors will be limiting or *where* a system will settle down. That depends on the specific nature of the factors and of the devices available to the system. If there *were* light bulbs whose lifetime increased with higher operating temperature, the problem of designing a long-lasting lighting system would not be the *same* problem, but it would always be a problem with multiple dimensions.

If ever a general systems law were important for our time, the Second Law of Regulatory Compromise is it. We live in an age blighted by the strategy of pushing things to their limits—and then the one step beyond. We have elevated the optimizers to the role of official government prophets, established specialism as a form of state religion, and begun to worship the idols of our own excesses.

Things weren't always this way. There are even a few isolated corners of the world where they still are not. The optimizers ridicule the rustics who inhabit these corners, and delight in pointing out how folk wisdom always contradicts itself. There is even a little intellectual game in which one player gives a folk saying, such as "haste makes waste," and the other replies with its contradiction, such as "a stitch in time saves nine." What they fail to realize is that the folk wisdom resides in neither dictum, but in the *acceptance of their mutual opposition*.

In one of these remote spots, in the Swiss Alps, an anthropologist once asked an old peasant how he defined the word "peasant." "A peasant," he said, after a short reflection, "is someone who does a little bit of everything." Could it be that the despondency and loneliness which characterize our age derive from the change from a culture where each person did "a little bit of everything" to a culture where each does "a lot of nothing?" If so, there is some reason to be hopeful for our future; people are specialists only because our machines are thus far so limited. With properly specialized machines, people could become generalists once more, leaving the specialist strategy where it belongs—with the machines. In other words, there is no reason why human beings in a society need *themselves* be specialists, just because the society as a whole adopts the specialist strategy of division of labor. On the contrary, it would be more prudent for them to remain as general as possible—to protect the society from specialist oversights.

Perhaps, then, when regulatory compromise becomes more widely understood, we can personally go back to a life following an even earlier dictum—"moderation in all things." To that, all our general systems laws can add is a paradoxical phrase:

"Moderation in all things—even in moderation."

QUESTIONS FOR FURTHER RESEARCH

1. *Pugilism*

A boxer preparing for a championship match spars with a partner who is heavier than his opponent—"because he hits harder"—and one who is lighter—"because he is faster." But though he may beat a slower opponent who hits harder and a lighter-hitting opponent who is faster than X, he may very well not be able to beat X. Explain.

2. *Military and Police Science*

The army in a modern state has at least two roles—to ensure domestic tranquility and to guard against the foreign enemy. Conflicts arise between these two roles when, for example, a foreign war is lost and people lose confidence in or respect for the army. Discuss the following:

1. The strategy of separation into distinct armies—"police" and "army"—versus a single policing force.
2. The strategy of separation of one army within the other—"national guard" within the army.
3. The suitability and possibility of arms supplied for one function—"foreign aid to protect against aggressors"—being used for the other—"suppression of armed revolt."

3. *Government*

Discuss the various current ideological systems of "government of the people" with respect to their relation to the First and Second Laws of Regulatory Compromise.

4. *Social Psychology*

The old adage says "there's safety in numbers." On the other hand, we are aware of situations in which bystanders refuse to act even when witnessing a slow and brutal murder. Social psychologists explain this lack of response partly in terms of the paradoxical influence of the crowd itself on its individual members, by which each member views the other's failure to act as evidence that the situation is not so serious after all: "If they're not alarmed by all that smoke, perhaps it's not a fire after all." In view of this nonindependence of a crowd of bystanders, discuss the limits to the "safety in numbers" strategy, according to the Laws of Regulatory Compromise and the Aggregate Laws. How does your argument depend on the nature of the community in which the action takes place, and what implications does it have for law enforcement policy? (Reference: Latane and Darley 1970)

5. *Writing*

In writing books, one encounters numerous problems of multidimensional regulation. For instance, a book may be used both as a reference and a text, and the two functions do not necessarily coincide. Another problem is the dual function of footnotes, to give credit to other scholars and to give information to readers. Discuss these and other multidimensional problems in writing a book; how they may be solved; what may result if they are not solved. Give examples from books you know or have written.

10

The Homeostatic Heuristics

Are there not general principles of stabilization? May not the devices developed in the animal organism for preserving steady states illustrate methods which are used, or which could be used, elsewhere? Would not a comparative study of stabilizing processes be suggestive? Might it not be useful to examine other forms of organization—industrial, domestic, or social—in the light of the organization of the body? (Cannon 1939:305)

In this chapter, we take up the challenge Walter Cannon so well laid down over half a century ago. Much has been learned about "preserving steady states" in the intervening years, but none of it has subtracted one word of truth from Cannon's original discussion. *The Wisdom of the Body* (1939), a true classic of systems thinking, can be read today with the same freshness of insight it brought to the world long before many of us were born. There's something about high-quality thought and writing that has a stability all its own, perhaps governed by some other process than explored in this chapter.

The Internal Environment

. . . the conditions of the organism and those of the surrounding environment seem to be independent; in these animals indeed the manifestation of vital phenomena no longer suffers the alternations and variations that the cosmic conditions display; and an inner force seems to join combat with these influences and in spite of them to maintain the vital forces in equilibrium. But fundamentally it is nothing of the sort; and the semblance depends simply on the fact that, by the more complete protective mechanism which we shall have occasion to study, the warm-blooded animal's internal environment comes less easily into equilibrium with the external cosmic environment. (Bernard 1957:62)

The "inner force" of which Bernard speaks is what we would call "structure"—that which stands between the input and the output, so that the system does not simply follow the environment's behavior. By providing a separate structure for each major variable in the "cosmic environment," the system frees each of its *other* subsystems from regulating that environmental variable for itself. This is the strategy of separation of environmental variables.

But while each mechanism is shielding the rest of the system from one component of the environment, it, in turn, may be shielded by some of the mechanisms it protects. In this way, no single part of the system is obliged to solve the full problem presented by the environment. Instead, each part operates in its own *internal environment,* a concept first clearly enunciated by Claude Bernard, the great French physiologist of the nineteenth century.

The internal environment, to Bernard, was a quite specific configuration in warm-blooded animals. To the general systems thinker, the concept is used more metaphorically, and broadly. "Internal" and "external" do not necessarily refer to the *physical* inside and outside of a system. "Internal" refers to those parts that are *under the protection of* the regulatory mechanisms, and thereby shielded from most of the slings and arrows of the "external" world, meaning, all that is not part of the system. A Swiss diplomat in Mexico is not geographically *inside* Switzerland, but is under the *protection* offered by the Swiss government to all its diplomats. The grizzly cub is protected inside the mother grizzly's body before birth, and after birth it still remains under her protection in many ways.

Still, the analogy is with a membrane, or skin, or sack, that surrounds the interior and protects it by interposing itself between the interior parts and the exterior threats. As Bernard observed, this protection is not *absolute,* but merely makes it "less easy" for the internal environment to come into equilibrium with the exterior. There will always be threats too big for any *particular* regulatory mechanism. If these are presented the system will not survive.

Carrying the sack analogy further, consider the use of an insulated bag in which ice cream can be carried home from the store. The insulation in the bag *retards* the flow of heat between ice cream and air long enough for us to get home before the ice cream melts. If the day is hot enough, or the trip is long enough, the insulated bag will fail to maintain the proper internal environment.

On the other hand, the bag protects the ice cream against other threats besides heat. Insects will have a difficult time in reaching the ice cream. So will children and dogs. Offensive odors may be kept out, especially if the bag is lined with plastic. From another point of view, the ice cream can be considered the threat, and the car seat the "internal environment" to be protected from food stains.

Although a membrane or skin is easy to imagine, there need be no such distinguishable boundary to create an internal environment. The outer layers of ice cream protect the inner layers from melting by sacrificing themselves to the heat. The outer leaves of a cabbage protect the inner ones from insects, but there is no definite place where "outer" stops and "inner" begins. People at the center of a sufficiently large mob are protected against police actions by the people at the periphery. This is why a large enough mob always overcomes the forces of law and order—at least before the days of long-range weapons and loudspeakers.

Actually, any mechanism that keeps some physical distance between the system and the threatening environment serves a regulatory function. In *compact* systems, like a mob or a cabbage, the space is provided by parts of the system itself. A skunk can fill its surrounding space with a material which many environmental threats do not care to penetrate. People can sometimes accomplish the same distancing with words, or symbols of power. Everyone knows that nobody is allowed to touch the monarch, even though nobody has ever tested the mechanism for this particular monarch.

The internal environment may be formed by something even more abstract than words. *Any* method that creates a lower "cost" to operating *within* the system than *into* the system from outside can effectively protect the interior from acts originating outside. Police with portable two-way radios can come quickly to each other's aid, perhaps faster than the threatening mob may be able to gang up on them. Without police radios, the mob may have the advantage of rapid communication, since they are compact and the police are spread out around their exterior, as illustrated in Figure 10.1.

In physical systems, *work* must be done to move *masses* over *distances*. Whenever a system is threatened by the movement of some mass from a distance, merely increasing the distance can create the equivalent of a protective skin. Inside the system, parts are close, and little work need be done to move masses from one part to another. Outside a remote system, the threats in the environment are far away, requiring much work to reach the system with any appreciable mass. Islands provide many varieties of plant and animal life with an internal environment free of potential competitors unable to afford the trip from the mainland.

In international trade, bulky goods of low monetary value cost much to ship across great distances between countries. This shipping cost must be added to the other costs of any foreign nation putting its goods in

Figure 10.1. Communication among the police is more difficult than within the mob, unless the police have portable two-way radios, which give them a superior "internal environment."

another nation's domestic market, shown in Figure 10.2. This shipping cost acts as an economic "barrier" tending to exclude certain goods which would otherwise create a much less hospitable "internal market" for the domestic firms. Such a barrier is just as much a protective mechanism as a set of tariffs and import quotas, though those under its protection like to speak of "free trade."

From this point of view, we can readily understand why the introduc-

Figure 10.2. Begonian goods are at a "natural" disadvantage on the Gloobian market because of the "barrier" created by shipping cost across Mungo.

tion of a new shipping technology *outside* the country can upset the internal equilibrium of trade *within* the country. The environment is no longer as "distant"—in terms of cost—so the internal environment is less effectively regulated. On the other hand, we can also understand how improved shipping *within* a country can upset foreign trade. Better roads bring California wines "closer" to New York, to the disadvantage of wines trying to penetrate the barrier from outside, as from France or Germany.

In short, the "internal" of "internal environment" is defined by the protection itself, not by its physical location, shape, or substance. Or, since protection is not simply automatic, we might say that the internal environment is that part of the system which is isolated from the external environment by one or more regulatory mechanisms.

A belief in the superiority of skin of a pinkish cast is surprisingly resistant to change. The resistance of this belief comes partly because it is protected by the skull from blows on the head, but not entirely. Most of its protection arises from its isolation from conflicting beliefs, and from an established set of behaviors keeping its owner from associations that might disabuse him of his racism.

Changing such deeply shielded beliefs is about as easy as changing one's body temperature by plunging into a cold lake. If the water is sufficiently cold to surpass the capacity of your regulatory mechanisms, the plunge may work, but then you are likely to die in the attempt.

Identifying and Essential Variables

The art of progress is to preserve order amid change, and to preserve change amid order. Life refuses to be embalmed alive.

Alfred North Whitehead

In a previous work (Weinberg 1975), we explored how the *identity* of a system was preserved. Indeed, we defined *survival* as retaining certain variables within limits set by some observer. From this point of view, the internal environment strategy is rather likely to protect a system's identity. If a system is to be seen, it must survive. If it is to survive, it must protect its identity. Thus the structure graph of such a system will likely show the identifying variables one or more steps *removed* from the direct influence of the environment, as shown in Figure 10.3.

Naturally, none of the system's variables can be *completely* isolated from the environment. Nevertheless, the identifying variables of a sur-

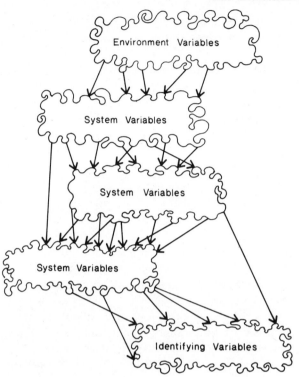

Figure 10.3. The identifying variables are one or more steps removed from the direct influence of the environment.

viving system are most likely to be accessible only indirectly, through action on other variables. This property, as Bernard and other physiologists knew, makes these identifying variables difficult to study. If you intrude on the body's interior environment, you are likely to derange the entire organism, not merely one variable. But if you try to influence one variable from the outside, the body resists with its exquisite system of defenses. The same mechanisms that prevent us from being poisoned also prevent us from being medicated.

Identifying variables will appear in close association with other variables. A marked change in one of these other closely associated variables will result rather quickly in a change in the identifying variables. Because these close variables are thus associated with the identity of the system, which is itself furthest removed from the external environment, they may be thought of as forming the internal environment—even if they are not actually found on the physical "inside" of the system.

In warm-blooded animals generally—and in people in particular—the pulse rate, body temperature, oxygen level of blood, respiration rate, and the number of bacteria in the tissues are all closely connected with one another—and with the maintenance of the set of variables by which we might identify a jaguar or a George. If, for example, George's respiration rate drops, his pulse rate may increase for a while. Eventually it will drop to zero if his respiration rate stays very long at zero. Soon after, George's body temperature will match the air temperature, and if George is not properly embalmed, bacteria will start to proliferate.

When a person joins a community, there will generally be some exchange of personal freedom of action in return for the regulatory services the community provides. You hang up your six-shooter. In return the sheriff will see that you don't have to worry about other people with guns. You get the convenience of regular garbage removal, but in return you must refrain from dropping your garbage wherever it happens to suit you.

In the same way, each part of a system that inhabits the internal environment has to sacrifice much of its independence of action in return. You may be able to survive in the desert if ridden out of Dodge City on a rail, but your kidneys won't last long if removed from the comfort of your gut. In strongly integrated systems like the mammalian body, the failure of one regulatory mechanism will generally lead to a sudden failure of the others, just like the collapse of strongly associated aggregates.

Our old-fashioned conception of "death" as a clearly defined event is probably based on this type of collapse. In ancient times, it was unknown for a person to survive without at least one functioning kidney for even a few minutes. Only with modern medical care taking over one regulatory function after another does it become difficult for everyone to tell when poor George is "dead."

Because this set of variables tends to stand or fall together, it is tempting to refer to them all as *essential* variables, as Ashby does in his *Introduction to Cybernetics* (1956).

When the set of variables is so interdependent, the question of which ones are actually the seat of system's identity becomes less important. One observer may view pulse rate as identifying; another, body temperature; a third, respiration. If any *one* of these essential variables fails, the failure of the others will follow so quickly that the three observers are unlikely to argue the question of survival. Only when a person is maintained by artificial means, such as in the intensive care unit of a hospital, is argument over the moment of death likely to become heated.

Not all systems, by any means, have such closely associated sets of variables. Indeed, it seems rather unlikely that any set of parts just thrown together into an arbitrary relationship could meaningfully be considered to have a set of *essential* variables. Only when the system has been carefully designed to have this property, or when it has evolved this property by sustained existence as a system in a taxing environment, can we successfully employ the "internal environment" metaphor. To the extent that the system was *not* so designed, or so evolved, the metaphor will be less successful.

In animal species, whether you prefer to believe in design or evolution, we find all the requirements for a set of variables to be called "essential." It isn't surprising, then, that a physiologist, Walter Cannon, coined the term "homeostasis," which means, roughly, "remaining the same." In the passage quoted at the beginning of this chapter, Cannon cautiously suggested that similar complexes of mutually regulating variables might be found elsewhere. He was right. Homeostasis and Cannon's observations about it have proved to be among our most general and fruitful heuristic devices.

The Homeostatic Principles

The great secret, known to internists and learned early in marriage by internists' wives, but still hidden from the general public, is that most things get better by themselves. Most things, in fact, are better by morning. (Thomas 1974:85)

The broad relationship among external environment, homeostatic mechanisms, and internal environment is summarized pictorially in Figure 10.4. We must add this picture to our answers to the question, "Who regulates the regulators?" This gives us three generalizations about regulation:

1. The First Aggregate Law.
2. The Second Aggregate Law.
3. The Internal Environment Principle.

In the context of physiology, Walter Cannon enunciated four subprinciples of the Internal Environment Principle, which he called *Proposition I, Proposition II, Proposition III,* and *Proposition IV.*

It may strike the reader that Cannon's names are not terribly memorable. Although we pay homage to his classical work, we think it would be a mistake to burden such useful principles with such useless names. Therefore, we shall take the liberty of renaming them, not out of disrespect for Cannon, but quite the contrary, out of respect for the enormous value of all four propositions as *heuristic* devices.

"Heuristic" is a useful word, but not familiar to everyone. It stems from

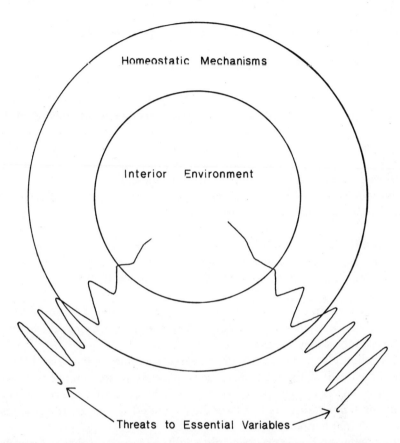

Figure 10.4. A schematic diagram of a homeostatic system shows how the interior environment is protected against a variety of exterior threats.

the same root as "Eureka"—"I have found it"—which Archimedes is supposed to have exclaimed as he leapt from his bath with the solution to the problem of the counterfeit crown. Cannon's propositions may not make you run through town wet and naked, but they will give you the same kind of flash of insight on a variety of problems if you use them as *heuristic* devices—ways of *helping you find out* what you want to know.

Of course, all of our general systems "laws" are heuristic devices, which is why we attempt to give them memorable and suggestive names. Cannon's first proposition, for example, we will call *The Pervasiveness Principle,* since it asserts, as we have emphasized, that regulation does not *just happen:*

Proposition I   *In an open system, such as our bodies represent, compounded of unstable material and subjected continually to disturbing conditions, constancy is in itself evidence that agencies are acting or ready to act, to maintain this constancy.* (Cannon 1939:299)

The Pervasiveness Principle establishes the point of departure we need when examining systems about which we have no previous knowledge. By probing an organization under this assumption, we must discover many of its workings. Many physiological secrets were unearthed by first considering what *problems* the body had to solve and only then searching for the organs which contributed to the *solution.* Guided by The Pervasiveness Principle, we will not allow ourselves to be content until we find how *each* essential variable is kept stable.

But what do we look for? Cannon's second proposition, which we dub *The Perversity Principle,* gives us our first clue:

Proposition II   *If a state remains steady it does so because any tendency towards change is automatically met by increased effectiveness of the factor or factors which resist the change.* (Cannon 1939:299)

In other words, in the search for the source of *constancy,* look for *activity.* The activity of regulatory mechanisms tends to be *contrary* to the activity of the variables they regulate.

A swollen gland is evidence of increased secretory activity, and therefore indicates that the gland may be responding to some regulatory call. Similarly, shivering is an increased activity when the exterior environment drops its temperature. It is an activity intended to reduce the response of the internal temperature.

Following the Perversity Principle might seem so easy that we wonder why physiologists still have so much to discover about how the human system regulates itself. Cannon's third proposition, gives us a clue to the source of adversity:

Proposition III   *The regulating system which determines a homeostatic state may comprise a number of cooperating factors brought into action at the same time or successively.* (Cannon 1939:300)

In other words, systems often carry the separation strategy further than we first imagined. Instead of a single mechanism for each variable, the system may possess a braid, weave, or *plait* of mechanisms, each adapted to some aspect of regulating that variable. To highlight the importance of disentangling this weave, we call Proposition III the *Plait Principle.*

Because of the Plait Principle, we cannot rest on our laurels when we have discovered one mechanism regulating a variable. To raise body temperature we not only shiver, but we tuck up into a ball, put on warm clothing, move about briskly, or build a fire. But Cannon's fourth proposition does give us some guidance in unraveling the plait:

Proposition IV *When a factor is known which can shift a homeostatic state in one direction it is reasonable to look for automatic control of that factor, or for a factor or factors having an opposing effect.* (Cannon 1939:300)

Actually, this statement embodies two heuristic principles, which we might call the *Pilot Principle:*

". . . automatic control of that factor . . ."

and the *Polarity Principle:*

". . . having an opposing effect."

These two principles are directly related to our Laws of Regulatory Compromise. Without *both* principles, as Cannon knew, the homeostatic mechanisms themselves would cause excursions outside the critical limits in one direction to be translated by overcompensation into excursions outside the limits in the other—as so often happens in diseased states. In a sense, then, these principles could be derived from the others, but our purpose is not to find a minimal set of heuristics, but a suggestive one. For that reason, we shall not only keep these two principles in our set, but shall, in what follows, search for other heuristics to add to the box of tools bequeathed to us by the physiologists.

QUESTIONS FOR FURTHER RESEARCH

1. *Organization of Writing Projects*
How is it possible for the set of problems for an entire chapter to get lost during production of a manuscript? How could a homeostatic system be designed to prevent such losses?

2. *Design of Problems*
Design a set of problems for this chapter. Work on those problems. What does a good problem have in common with a homeostatic system? What does a good *set* of problems have in common with a homeostatic system?

11

Other Regulatory Heuristics

But if the part acted on by centrifugal force, instead of acting directly on the machine, sets in motion a contrivance which continually increases the resistance as long as the velocity is above its normal value, and reverses its action when the velocity is below that value, the governor will bring the velocity to the same normal value whatever variation (within the working limits of the machine) be made in the driving power or the resistance.

I propose at present, without entering into any details of mechanism, to direct the attention of engineers and mathematicians to the dynamical theory of such governors. (Maxwell 1868)

General principles of regulation have been discovered and rediscovered over the ages, but their application has until recently been within the narrow field of their discovery. These principles are not the intellectual property of any one field or any one person but are the common heritage of all of us. Only by studying them in their full generality can we make them available on a democratic basis to everyone, regardless of discipline, vocabulary, or mathematical training.

The Feedback Principle

Feedback is a method of controlling a system by reinserting into it the results of its past performance. (Wiener 1954)

Although Maxwell had laid down the fundamental principles of "governors" 100 years earlier and Cannon had set down numerous physiological examples, Norbert Wiener is popularly regarded as the "father of cybernetics." He did, certainly, recoin the word in the modern usage, from the

Greek word for "steering" or "helmsman." He also developed a great deal of mathematical machinery describing the behavior of various cybernetic mechanisms—(see Willins 1971)—"agencies acting to maintain constancy."

Weiner and his followers studied both natural and artificial systems of regulation. Perhaps because people accept the idea that the human body regulates such variables as temperature, the idea of "agencies acting to maintain constancy" usually includes a vitalistic element. How inanimate agencies—and particularly unthinking agencies—could act to maintain constancy is a mystery, until some simple mechanisms are explained.

Usually in discussions of mechanical controllers, or "servo-mechanisms," the nontechnical reader is put off by machines themselves. Authors try to help by using familiar systems. The household thermostat takes the palm in this competition. But many people live in apartments without thermostats. Even for those people who do have one, a thermostat is difficult to understand because part of it is in one place, part of it is in another, and invisible wires connect the two parts. Therefore, we shall begin with a different example—one that is ubiquitous, inconspicuous, and ignominious—the flush toilet. (See Figure 11.1.)

The flush toilet is not ordinarily thought of as a regulatory device for the same reason we ignore many other regulators: it works too well. But if you have a tank toilet, you may observe its white box operation by re-

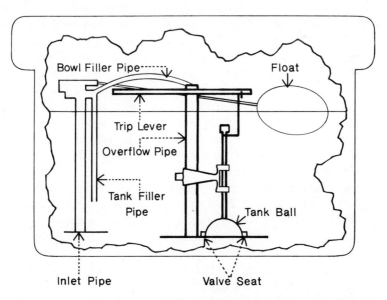

Figure 11.1. The ordinary flush toilet of the United States *circa* 1975.

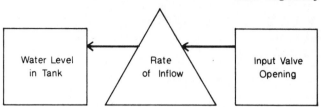

Figure 11.2. The mechanism of filling the tank consists of opening the valve on the tank filler pipe to allow water to flow into the tank.

moving the top of the *tank*—not the stool. (Don't worry—that's the *input* water in there, not the output water. Any discoloration is due to other people's sewage, not your own.)

Now that you have overcome the prohibitions of a puritanical upbringing and have the white box open, you can see that the water is standing rather high in the tank, waiting for the next flushing. Now flush the toilet and watch how the water goes out of the tank, dropping the level almost to the bottom. That part of the operation is easy to understand; but now some mechanism must fill the tank to the operating level and keep it there until the next flush. You see that it is happening, but how?

We know from the Bathtub Law that the level of the water at any moment is determined by the history of inflow and outflow. Since the Bathtub Law is a general systems law, it applies to toilet tanks as well as to bathtubs, and since the outflow has been stopped immediately after flushing ($D = 0$), the available mechanism for regulating the water level is the inflow, B. The rate of inflow is determined by the size of the opening of a valve, shown in Figure 11.2. Notice how we use a triangular block to indicate a *rate*, so we won't confuse it with an *amount*. The distinction between rates and amounts is a crude, but often useful, way to get a rough idea of the time sequence of behavior of a continuous system. The amounts tend to determine the short-range behavior (the initial conditions) and the rates tend to take over determining the behavior as time goes on (the equifinal condition).

In the diagram of Figure 11.2 there is no indication of feedback, no way of "reinserting the results of past performance." Something additional will be needed to convert this diagram into a regulatory system. This something else *could* be a machine, but suppose for the moment we don't have a machine. Could we do the job ourselves?

Because the Bathtub Law applies, we know just what to do. We can control the filling of the tank in the same way we control the filling of the tub, by opening a spigot and letting the water run in until we see

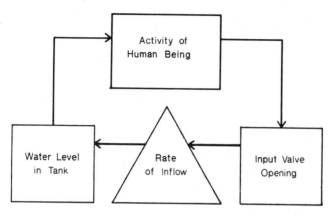

Figure 11.3. A human being can regulate the water level in the tank by opening the valve whenever the water level in the tank is below the desired level.

that the level is where we want it. This new system is shown schematically in Figure 11.3.

Because we are performing part of this job ourselves, we have no trouble believing in the power of this system to regulate. We don't have to try it out on an actual toilet to believe it, because we've often regulated the level of water in a sink or tub the same way. With a little foresight, we can even regulate the water level in the absence of the usual water supply. When the plumber disrupts our regulatory activities by shutting off the water, we need only be prepared with buckets and a swimming pool to maintain our system. Yet if we can replace each of the steps we performed ourselves by a perfectly mechanical system, we will demonstrate that there is nothing in the regulation that is intrinsically human or sentient.

Analyzing Feedback Loops

The use of feedback in systems may not necessarily improve performance. In fact, if the goals are not clearly understood, it may create more problems than it solves. (Cruz 1972:2)

Artificial feedback systems, at least, can be analyzed to high orders of mathematical precision, but learning to perform such calculations is not essential to an understanding of feedback concepts. Indeed, many engineers get so lost in the mathematical details of feedback analysis that

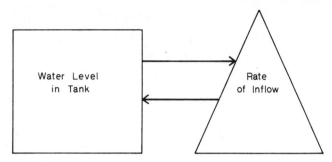

Figure 11.4. The essential portion of the diagram of possible effects shows that the water level in the tank affects, and is affected by, the rate of inflow.

they lose sight of the goals of the system and the goals of the analysis itself. Our analysis will be at such a general level that no such shortsightedness should be possible.

Before we mechanize each of the steps, let's trace out the underlying secret of the regulation in the human system. First we draw the transitive closure of Figure 11.3, giving Figure 11.4, in which all but two essential boxes have been eliminated. Removing the other boxes emphasizes the central idea of feedback regulation. The water level in the tank ultimately influences the rate of flow of water into the *same* tank, which then influences the level of water. In this way, "past performance"—the level of water—is inserted into the process that produced that performance, the rate of input flow. This is "feedback"—the essential element in a servo-mechanism.

Although feedback is essential for a servo-mechanism, we would be mistaken to conclude that *every* feedback loop leads to stability. In an earlier example (redrawn as Figure 11.5) there was a feedback loop between the 0–14 and 15–45 aggregates, because the 15–45 group "fed back" births into the 0–14 group. Yet this particular loop led to a "runaway," or exponentially growing, population. Conversely, with a different set of

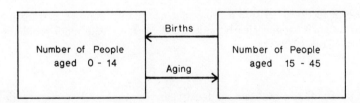

Figure 11.5. Feedback from the adult population to the child population by way of births could lead to "runaway," rather than stability, showing that feedback is not sufficient for stability.

structural values, the loop would have led to "extinction," or "runaway to zero."

Much has been made, in the cybernetic literature, of the difference between "runaway to infinity" and "runaway to zero," but the distinction is not terribly productive. We can always reverse what we call "infinity" and what we call "zero." The Principle of Indifference tells us that the distinction is artificial. It's merely a matter of convenience in naming things.

For instance, suppose we weren't interested in "number of people," but in "average space per person," as shown in Figure 11.6. If we had happened to choose these variables, what was previously seen as a "runaway to infinity" of people would now be seen as a "runaway to zero" of space per person. Similarly, the "runaway to zero" people would become a "runaway to infinity" of space per person.

We know that the conclusions we reach from analyzing a diagram or equation *should not* depend on the way we label our variables. We also know, from the Principle of Difference, that our minds often function better with a more suitable set of names, so that we prefer some ways of naming things over others. If we are careful, explicit, and consistent in our labeling, however, we need not get into trouble. Moreover, if we do it properly, we should be able to deduce from a structure diagram whether or not there is a *potential* for stability.

To look for potentially stabilizing feedback loops, we start by putting a plus or minus sign on each arrow of the diagram. A *plus* sign is used if an *increase* in the "from" variable tends to cause an *increase* in the "to" variable—when "from" goes up, "to" goes up. A *minus* sign is used when an *increase* in the "from" variable tends to cause a *decrease* in the "to" variable—when "from" goes up, "to" goes *down*.

In Figure 11.7, we've illustrated this plus and minus marking with our toilet regulator. Consider the arrow *from* "water level" *to* "human activity." The activity here is "closing a valve." Does that increase or decrease as "water level" is increased? As we see the water get higher, we start to

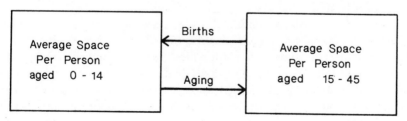

Figure 11.6. The same system of Figure 11.5 can be described in terms of different variables—the "average space per person"—making the "runaway to infinity" become "runaway to zero."

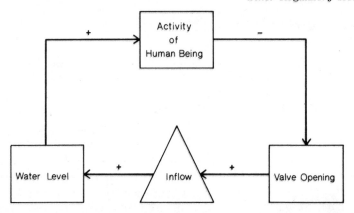

Figure 11.7.   Arrows on the diagram of effects can be labelled to indicate the direction of influence of one box on another.

close the valve, to prevent overflow, so *increasing* "water level" leads to *increasing* our activity of valve closing. Therefore, we put a *plus* on this arrow.

Careful reading of the "from" and "to" influences prevents any mistakes in marking the arrows. If the "human activity" had been expressed as "opening the valve," the opposite of what we used, then *increasing* water level would have *decreased* human activity—that is, led to less opening of the valve.

In the actual diagram, the variable was "closing the valve." How did this variable, in turn, affect the variable "valve opening"? Since *more closing activity* leads to a *less open valve*, this second arrow is labeled with a *minus*. Notice, though, that if we *had* used "opening the valve" as the "human activity," this label, also, would have been changed.

How are the rest of the labels obtained? An *increased* opening of the valve would *increase* the rate of inflow, so we put a *plus* on this arrow. Finally, an *increased* rate of inflow would *increase* the water level, so another *plus* is used.

With the entire diagram marked, we can use these pluses and minuses to mark directions of influence on the transitive closure, or diagram of possible effects, as shown in Figure 11.8. This marking is accomplished by combining the markings on the original arrows one pair at a time, as each pair of arrows is condensed into one arrow on the possible effects diagram. If both original arrows are the same sign—both plus or both minus—the "effects" arrow is given a plus sign. Why?

Consider three variables, *A*, *B*, and *C*, where *A* affects *B* and *B*, in turn, affects, *C*. If an increase in *A* leads to an increase in *B*, and an in-

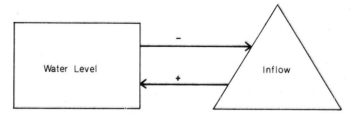

Figure 11.8. The labelling of direction of influence can be carried forward on the diagram of possible effects to give an indication of the direction of secondary influences. In this case, the labelling shows there is potentially negative feedback around the loop.

crease in *B* leads to an increase in *C*, then, clearly, an increase in *A* ought to *ultimately* lead to an increase in *C*, through its action on *B*. The two original plus arrows thus combine into a plus arrow from *A* to *C*. On the other hand, if an increase in *A* leads to a *decrease* in *B*, and an increase in *B* leads to a *decrease* in *C*, then *A* increasing will tend to drive down *B*, which in turn will drive *up C*. That's why the minus from *A* to *B* and the minus from *B* to *C* combine to make a *plus* from *A* to *C*.

Please note that this reasoning is extremely *qualitative*. In any actual case, we may not be able to draw such conclusions. For instance, although increasing *B* might decrease *C*, decreasing *B* might *not* increase *C*. *C*, for instance, might be a "one-way," or "ratchet" variable. Or, the speed of these effects might not be great enough to make it reasonable to state that *A* influences *C* at all, on the time scale needed to make an effective feedback loop. Finally, not all variables can be so easily characterized by simple effects. Increasing *B* might decrease *C* under some circumstances and increase it under others, perhaps depending on where *C* lay in its range of possible values.

Nevertheless, the *converse* reasoning can be quite powerful. That is, we may often be able to reason correctly that no stable feedback loop can exist with these variables, and that will tell us to widen our search for the source of homeostatic behavior. How is this done?

First, we complete our labeling of the diagram of possible effects by applying a minus sign to the closure arrow whenever the two original arrows have one plus and one minus. When we have finished, we look at the block we are trying to regulate and see if it has *opposite signs* going in and out of some other block, as we have seen in Figure 11.8. If the signs are opposite, there is a *chance* of stability. If they are not opposite, a runaway to somewhere is the inevitable result.

As a check on the process, we show in Figure 11.9 how the diagram of immediate effects would have been labelled if instead of "water level"

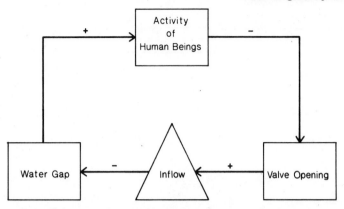

Figure 11.9. When we change the variables in the diagram, we must be careful to change the labelling of the arrows accordingly.

we had chosen "gap between desired and actual water level" as one of our variables, thus reversing the direction of increase. By labeling carefully, taking these directions into account, we wind up with the same sort of diagram of ultimate effects in Figure 11.10 as we got in Figure 11.8. And we *should* get the same conclusion, by the Principle of Indifference.

This stability property arises because, as Cannon said, "tendency towards change is met by increased effectiveness of the factors which resist the change"—the Pervasiveness Principle. As the signs in Figure 11.10 show us, a tendency to increase water level past the desired point is met by a decrease in inflow, or, alternatively, water below the level leads to increased effectiveness of inflow, which tends to restore the level. But *how* is the change met? Just what role does the human being play?

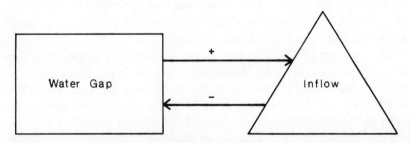

Figure 11.10. Diagram of ultimate effects derived from Figure 11.9. This diagram is equivalent to the diagram of Figure 11.8, even though the labelling is different. Both show that there is negative feedback around the loop.

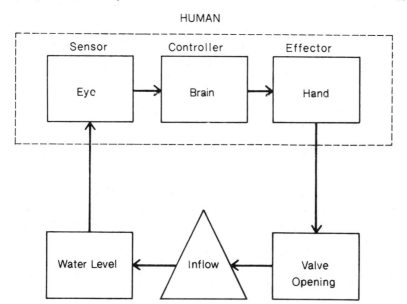

Figure 11.11. Components of the human servo-mechanism for refilling the toilet tank.

As Figure 11.11 shows, the person must do three things:

1. Sense whether or not the water is at the proper level.
2. Effect the opening and closing of the valve.
3. Control the process by correlating 1 and 2 to open the valve when the water is not at the proper level and to close the valve when it is.

In the language of servo-mechanisms, 1. is the "sensor" activity, 2. is the "effector," and 3. is the "controller." The "proper level," mentioned in 3., is called the "set point," because in an automatic device that is where we set the mechanism to do what we want.

Now, *any* device that can perform these three functions can do the same job the human being does (Figure 11.12). The usual device for flush toilets is a simple float (the sensor) resting on top of the water. As the water rises, the mechanical connection (the controller) between the float and a stopper (the effector) causes the stopper to move and closes the valve. If the connection has been adjusted properly (the set point), the valve closes completely just as the float is raised to the desired level. There it rests until the next flushing, when the float drops again, opening the valve and starting the *automatic* refill of the tank. Try it and see.

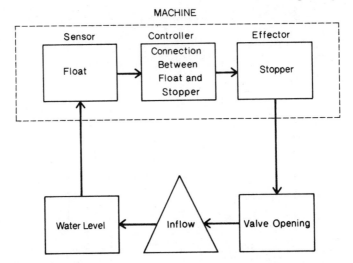

Figure 11.12. Components of the mechanical servo-mechanism that is equivalent in function to the human servo-mechanism of Figure 11.11.

The Piddling Principle

At this moment the King, who had been for some time busily writing in his note-book, called out "Silence!" and read out from his book. "Rule Forty-two. *All persons more than a mile high to leave the court.*"

> Everybody looked at Alice.
> "*I'm* not a mile high," said Alice.
> "You are," said the King.
> "Nearly two miles high," added the Queen.
> "Well, I shan't go, at any rate," said Alice; "besides, that's not a regular rule: you invented it just now."
> "It's the oldest rule in the book," said the King.
> "Then it ought to be Number One," said Alice.

> Lewis Carroll
> Alice in Wonderland

Cannon's Homeostatic Principles and the Feedback Principle are powerful general instruments of systems thought, but do they say *all* that is worth saying in general about regulation? We think not. We intend to continue our quest for other usable principles. In doing so, we must not be discouraged by the very success of the ones we have already at hand.

The sophisticated servo-systems built under the guidance of cybernetic theory have brought deserved acclaim to the cyberneticists. Like any public success in American society, cybernetics has seen its name become not just a household word but a marketable commodity. Popping up everywhere like mushrooms after a spring downpour are such neologisms as "psychocybernetics," "cybernation," "cybercultural," and "cyberarchal." Leaving aside the damage done to the Greek and English languages, the net result of all this talk is probably harmless. There is, however, at least one danger to systems thinking that can be directly attributed to the popular success of cybernetics. That danger is *undergeneralization*.

For example, Wiener's original work (1948) was on the automatic aiming of guns, and the concept of "aim" or "goal" or "purpose" has been frozen into the very name of cybernetics ever since. This purpose is to carry out "commands" which take the form of a set point at which the level of the variable to be maintained is determined from outside the system. But the concept of an aim, and an aimer, or steerer, or helmsman, is something which we do not find in, say, Cannon's work. Cannon speaks of "constancy"—not of being "on target," or "searching for the goal."

For the general systems thinker, then, feedback *control* systems represent only a subclass of feedback systems, since we do not wish to rule out the possibility of a system in which no explicit set point or "controller" can be found. Consider a system such as modeled in Figure 11.13. Where is the set point in such a system? Indeed, which is the variable

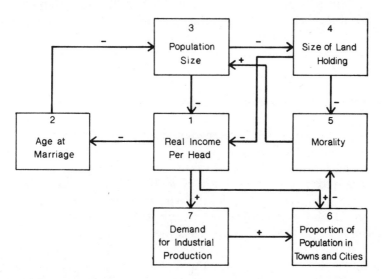

Figure 11.13. A hypothetical model of relationships between demographic, social, and economic change (Wrigley 1969: 109).

being "controlled?" It is not only a question of "who regulates the regulators?" but also of "who is regulating what?" Is age at marriage being regulated by population size? Or is population size being regulated by age at marriage? Or are both being regulated by real income per person?

The fact of the matter is that in such a system, as in a human body, each of the variables regulates for each of the others. The set point idea may be very appealing, especially to harried administrators, but certain systems just don't fit that image. The danger to systems thinking is that we may begin to *see* the world in our own cybernetic image; the danger to us all is that bureaucrats may begin to *make* our world in that image. Undergeneralized thinking can soon lead to undergeneralized living.

A second undergeneralization danger is in limiting our thinking to the concept of feedback itself. Wiener himself was under no illusions about this, for he said "feedback is *a* method of controlling systems . . ." not "feedback is *the* method of controlling systems . . ." It is an interesting comment on the intellectual herd instinct to note that, before Wiener, people had difficulty imagining feedback regulation, while after him, they have difficulty imagining any other sort.

Are there other principles besides the Feedback Principle? We know there are. We have already examined the strategy of regulation by sheer force of numbers, which we might call the *Ponderous Principle*. But open systems have other strategies, among the most interesting of which is not staying big, but staying small. We frequently find systems that regulate some variable by associating with another system that is much bigger and has its own mechanism for regulating. The flush toilet, for example, depends on a reliable outside water supply. The nature of that mechanism need not concern the smaller system unless it begins to grow larger. In other words,

When regulating by association, it is important to be unimportant.

For this reason, we may call this method the *Piddling Principle.*

We are surrounded by examples of systems applying the Piddling Principle. In biology, we have parasites, such as tapeworms, that depend on their host for nourishment—but they must not take too much, lest they end a good thing. We can model the parasitic relation with the digraph of Figure 11.14, whose labels show clearly that the parasite and host are locked in a stable feedback loop. Thus the success of parasitism is self-limiting—though there is no explicit set point establishing the limit. From the point of view of the parasite, this stability is bad; from the point of view of the host, it places a limit on misery.

But the existence of this particular feedback loop should not blind us

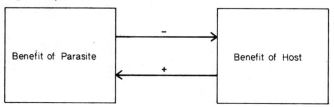

Figure 11.14. *Simple parasitism* is shown to be self-limiting, because too much benefit to the parasite has too negative an effect on the host, which eventually acts against the benefit of the parasite.

to the necessity, for the parasite, of the host being a good regulator in its own right. In fact, the parasite may benefit in the long term from the host's regulatory mechanisms directed against it, as demonstrated in Figure 11.15. When the parasite first invades the host, many of its siblings may be invading at the same time. If the host cannot reject most of them through its mechanisms for resistance, the entire group of parasites will be too successful, and the Piddling Principle will be violated. But if the

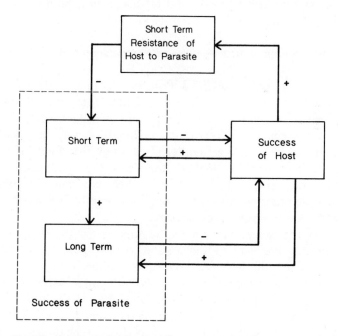

Figure 11.15. The parasite's long-term success depends ultimately on the host developing a certain amount of resistance to the parasite in the short run, so that the host is not killed by the parasite.

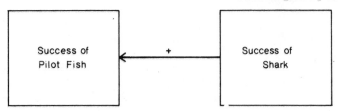

Figure 11.16. The *scavenger strategy* requires tha' the shark be prosperous and so produce a lot of garbage for the pilot fish to live on.

host succeeds—though not entirely—in rejecting parasites, then the long-term success of the survivors is put on a more secure basis.

Perhaps the strategy of regulation through association is more easily seen in the case of *scavenger systems,* which live off the waste products of some larger system. For example, there is the tiny pilot fish, which is small enough to live off the shark's leavings and also too small for the shark to profit significantly from eating it. In its simplest form, the scavenger strategy is illustrated in Figure 11.16, without any feedback whatsoever in regulating the success of the pilot fish.

Yet, as we dig deeper, we do find feedback—of a positive kind. It is a general rule in nature that

No system can long endure by eating its own excrement.

We might call this the *Foodback Principle,* to remind us that though a living system may seem to be in a closed input/output relationship with itself, some hidden input must exist, if only to make up losses. If a bacterial culture seems to be existing indefinitely merely by cannibalizing the remains of its earlier generations, we know that food in some form must be leaking in. With "higher" organisms, the Foodback Principle is even more powerful a limit, for by "higher" we generally mean more dependent on other organisms to provide food and other services. One of those services is the removal of garbage.

In a very real sense, the "higher" organisms are the sloppier, and no scavenger can live without the assistance of a slob. As a system gets more successful, and thereby increases its rate of garbage production, it is faced with a self-limiting accumulation of its own excrement, as shown by the feedback loop in the dark lines of Figure 11.17. By introducing a scavenger, however, the accumulation of refuse is decoupled to some extent from the rate of production, and to that extent the scavenger pays for its dinner by helping the slob.

In the end, then, almost any scavenger may be seen to be in a *symbiotic*

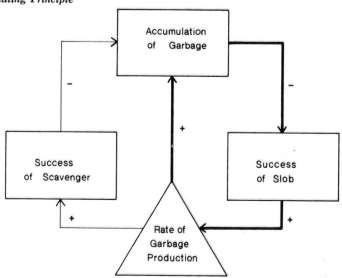

Figure 11.17. How the scavenger performs a service to the slob whose garbage it lives upon, by preventing a fatal accumulation of garbage.

relationship with its slob (Figure 11.18). The pilot fish actually serves the shark by playing dentist to the shark's many teeth. But, though the positive feedback loop of symbiosis might seem to augur for unlimited success for the scavenger, what the diagram cannot show is that the scavenger is ordinarily limited to being much smaller than its slob. Tramps, hobos, beggars, and other vagabonds perform a valuable service to society, but not everybody can make a living by being a bum.

In economics, the Piddling Principle is expressed in the idea of the "perfect market"—one in which the participating individuals are too small for any one of them to affect prices in any significant way. In sociology, the same idea is expressed in what Slater (1970) calls "the toilet assumption"—the idea that anything we don't need can be flushed away through

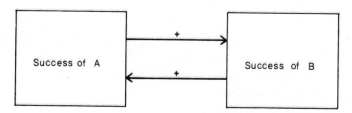

Figure 11.18. The *ultimate effect* of symbiosis must be mutually reinforcing for both parties.

that universal regulating device, the toilet. The problem with the toilet assumption as an operational principle is that it is usually not coupled with an appreciation of the Foodback Principle. When Alice grew to be 9 feet tall, it made her unhappy and she cried. When she shrank to 3 inches, her tears seemed like a lake. If we persist in making the world our toilet, we too, shall swim in tears—or worse. *That* is Rule Number One.

The Parallel Principle

During the first stage in the development of studies in economic history, each sector of economic life was treated separately, thus producing histories of agriculture, commerce, industry, banking, and so forth. In the second stage, the separate histories were joined on a national scale, but still not integrated. . . . We now know that a change in one of the segments causes repercussions in many other parts of the economy and the social structure. The ideal equilibrium of the old economic theories is constantly disturbed. From a historical point of view, the disturbing factors are particularly interesting. (Slicher Van Bath 1968:604)

Cyberneticists often classify all nonfeedback methods of regulation as "programmed," or nonvarying mechanisms, but the term was coined before the tumultuous growth of modern digital computers. Now, "programmed" no longer necessarily means something immutable, and it may even imply astonishing versatility. Moreover, *all* systems are programmed, in the sense of having some relatively fixed part that plays a role in determining behavior. The feedback system's behavior is not entirely determined by the feedback information, but also by the program, the structure of the system. Information being fed back need not mean the system will regulate at all.

For example, a classic and simple feedback situation is the relationship between the price of a commodity and the amount produced—the famous "Law of Supply and Demand." In Figure 11.19 we see the workings of this system reduced to its simplest form:

1. An increase in price fetches out increased production (a plus on the arrow from price to production).
2. An increase in production drives down the price (a minus on the production-price arrow).

But in Figure 11.20, we see a similar situation with entirely different results. In at least some circumstances, the price of *labor* (wages) has a

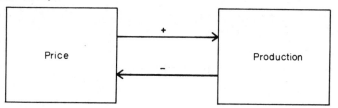

Figure 11.19. The *Law of Supply and Demand* shows that prices and production act as mutual regulators.

negative influence on the production of new *laborers* (births). In simple terms, "the rich get rich and the poor have children." In both these figures, information on current prices is fed back to producers, but the first leads to stability and the other, with a slightly different program, leads to runaway.

Since all systems thus have programs, it does not seem productive to distinguish between "programmed" and "feedback" systems. In the same way, it turns out that essentially all systems that survive have some form of feedback—some way in which information from the past is brought to bear on the behavior of the present. Put another way, all feedback systems are based on the Axiom of Experience: "The future will be like the past . . ."

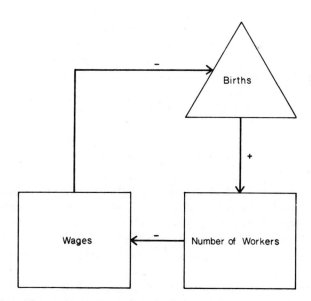

Figure 11.20. The supply and demand law doesn't apply for labor in the long run if raising wages lowers the birth rate.

Were there no continuity between past and future, no feedback system would work at all. If the valve opening sometimes brought in water and sometimes let it out, if water appeared and disappeared without obeying even the most rudimentary conservation law, if the float sometimes floated and sometimes sank, willy-nilly—in the face of any or all these discontinuities, even a simple flush toilet would not be possible, not to speak of life itself. By the Structure-Regulation Law, all regulation rests on the success of some other regulation, at least back to some of the deepest laws of nature.

The importance of continuity with the past is easiest to see in the feedback regulators, for if everything changes from one moment to the next, what is the use of information about previous behavior? If the roulette wheel is really random, one need not bother asking whether the red or the black just won. In that situation, one bet is as good as another. Similarly, if we try to shoot down an airplane with a servo-guided missile, we shall have no success whatsoever if the airplane can simply move from one place to another in an instant without passing through the intermediate positions. Only because continuity is forced on the plane by the laws of physics can we shoot it down—provided our missile is at least as maneuverable as the target.

From such considerations, we begin to get an inkling of perhaps the most powerful heuristic of all for studying regulation. Cannon hinted at it in the Pervasiveness Principle, when he observed that "constancy is in itself evidence that agencies are acting or ready to act . . ." and in the Perversity Principle when he spoke of tendencies to change being met by factors to resist change. Sometimes, if we are sufficiently clever, we can find much more than these principles hint. Because the regulatory mechanisms are there to meet disturbances from the environment, and because they must counteract these disturbances, they must, in some sense, obey the *Parallel Principle:*

A regulator must be "like" the environment it regulates.

The guided missile must be like the airplane in the sense of being at least as maneuverable; the flush toilet has a mechanism that operates intermittently, like the use of the toilet itself. If the environment is fast, the regulator must be just a little faster; if the environment is big, the regulator must be a little bit bigger; if the environment is full of surprises, the regulator must be full of tricks to counter them.

Even social systems exhibit the Parallel Principle. In the French village of Wissous (R. T. and G. Anderson 1962), a proliferation of voluntary associations during the last 50 years has enabled the community to adapt to urbanization without losing its identity. Before 1920, Wissous was an agricultural village of less than 1000 people—close to Paris but isolated from the capital by cultural distance. The tide of urbanization began to sweep the village in the 1920s. With improved transportation and communication came an increase in population. By 1962, of the 2500 residents of Wissous, only 200 were still farmers, while many of the others worked in Paris and only slept in the village. The four village associations of the pre-1920 era have multiplied into 40. They represent a bureaucratization of traditional village activities and institutions that can no longer be managed informally. This "replicate social structure" has important regulatory functions:

On the local level, the president of an association can meet with the head of any other local unit, including the village mayor, to speak authoritatively for his group. Beyond the village, he may speak for his group directly. But more effectively, he may transfer his responsibilities to the head of the regional or national association, so that the leadership of larger non-village bodies can face association representatives on the same socio-political level. With power inequalities thus minimized, the association fits easily into both the local and the national structure. (R. T. and G. Anderson 1962:368)

The village of Wissous has become parallel in structure to the French state by conscious imitation, but for the same reasons that any system tends to become "like" the environment it regulates. The replication of traditional village functions in bureaucratic form "eases social life in a modern society by permitting these groups to articulate with one another and with a modern bureaucracy on all levels" (R. T. and G. Anderson 1962: 369).

All long-term regulators are programmed, consciously or unconsciously. The "design" of their program matches the "design" of the environment, for were that not so, the regulator would fail to regulate. Failing to regulate, it would pass from existence, and we would not see it.

With such selection against poor regulators, information from the past—perhaps the far distant past—feeds back into the design of the regulatory mechanisms that survive. Some programs are simple. Their past environments were probably simple. Some programs are elaborate. Presumably they came from an elaborate past, even though their present may not require such elaborate programs.

An overly elaborate program can survive, especially if some parts of the program are never invoked. A baby can swim, though few modern babies need that ability to get through the first year of life. But though a regulator can be far *more* capable than needed in the present environment, it won't last long as a regulator if it's far *less* so.

Thus it is that, with appropriate cautions, we may be able to read a system's past from its present regulatory mechanisms. Thus it is, also, that knowing a system's past, we may know what to look for in the way of regulatory mechanisms. As Slicher Van Bath says in the quotation that begins this section, "From a historical point of view, the disturbing factors are particularly interesting," because the present system is like the past environment.

Whether we can use the Parallel Principle all depends on our agility in defining what "like" means when we say that a system will be "like" its past environments. Learning to use the Parallel Principle is learning to twist out the "correct" definition of "like" for each system we happen to study. But general systems thinking is always that way, based as it is on a mathematics of metaphors, an algebra of analogies. In the next few chapters, we'll play some games with "like," to see what insights we can fetch out of the Parallel Principle.

QUESTIONS FOR FURTHER RESEARCH

1. *Sociology*

One gets a clue to the nature of elementary collective behavior by recognizing the form of social interaction that has been called *circular reaction*. This refers to a type of interstimulation wherein the response of one individual reproduces the stimulation that has come from another individual and in being reflected back to this individual reinforces the stimulation. Thus the interstimulation assumes a circular form in which individuals reflect one another's states of feeling and in so doing intensify this feeling. It is well evidenced in the transmission of feelings and moods among people who are in a state of excitement. One sees the process clearly amidst cattle in a state of alarm. . . . (Blumer 1946:170)

Diagram circular reaction with a plus-minus labeled digraph, then elaborate the diagram with mechanisms for a specific case, such as a lynch mob or a stampede of cattle. Show on your diagram how additional mechanisms could be added to the situation to regulate the collective behavior and prevent collapse or runaway.

2. *Anthropology*

Ethnologists tell of a tribe of Indians in the Southwest who were so poor that in bad years they would return in autumn to the place where in the spring they had gorged themselves on the abundant fruit of the yucca plant, to pick the seeds out of their own excrement in order to stay alive—the so-called "second harvest of the yucca." Does this behavior violate the Foodback Principle? Explain, and give other examples of a system seeming to live off its own remains.

3. *Plumbing and History*

The float valve regulator of the type found in the flush toilet seems to date back at least to the first century A.D. (Heron of Alexandria), or perhaps even earlier. As a regulator of flush toilets, it dates back at least to 1740, in England. Yet there are many other regulatory mechanisms involved in the disposal of human wastes, some feedback and some not. Their importance in history may be inferred from the historical importance of plagues and from the observation that during World War II, intensive bombing of cities never seemed to destroy a city unless it destroyed the sewage system. Trace the history of plumbing and sewage, from the point of view of a regulatory system for eliminating foodback. (Reference: Mayr 1970)

4. *Lichenology*

The prime model of symbiosis is the paired-system plant we call lichen. Discuss the strategy of the lichen, how it may have evolved, and variations in the structure of the symbiotic relationship between the fungus and the alga. In particular, are there any three-way symbioses involving lichens? (Reference: Hale 1970)

5. *Medicine and Dentistry*

Discuss the following provocative statements: "Dentists are parasites living on decayed teeth," and "The path to improved medical care is to pay the doctors for the health of their patients, not for their sickness."

6. *Medicine, Dentistry, and all other Professions*

Discuss the following in terms of regulatory strategy:

"The three basic principles of professional ethics are:

1. Strict sanctions against price cutting in any form.
2. Strict regulation of entry into the profession.
3. A tacit agreement to cover up the mistakes of the practitioners by a conspiracy of silence directed toward preventing any feedback to the public." (Boulding 1956)

In particular, do not confine yourself to nodding your agreement about *other* professions, while shaking your head violently about your own.

7. *Football and Other Sports*

Discuss the Parallel Principle applied to the offense and defense in sports such as football, basketball, baseball, and hockey. In particular, discuss how it explains the "surprise" which always seems to result when two teams from different and previously isolated traditions first meet—as in the third Super Bowl or the first meeting of the Russian and Canadian national hockey teams, each of which resulted in "upsets."

8. *Incarceration*

Social scientists have observed that long-term prisoners come to behave like their jailors—for example, in Nazi concentration camps during World War II. Discuss this phenomenon in light of the Parallel Principle. Develop other examples in society, such as students and professors, subordinates and bosses.

9. *Parallel Lifetimes*

. . . nothing preconceived in what we call harmony in Nature. The chance of collisions and encounters has sufficed to establish it. Such a phenomenon will last for centuries because the adaptation, the equilibrium it represents has taken centuries to be established; while such another will last but an instant if that form of momentary equilibrium was born in an instant. If the planets of our solar system do not collide with one another and do not destroy one another every day, if they last millions of years, it is because they represent an equilibrium that has taken millions of centuries to establish as a resultant of millions of blind forces. If continents are not continually destroyed by volcanic shocks, it is because they have taken thousands and thousands of centuries to build up, molecule by molecule, and to take their present shape. (Kropotkin 1970:120–121).

Discuss the social form of this principle of parallel lifetimes. For example, in predicting how much longer a marriage is going to last, is the "best" prediction somehow the length of time it has already lasted without divorce? Is the length of time a business will last somehow predicted by the length of time it has already lasted? How about a religion? A government? A peace? A war?

10. *Symbiosis*

The honey badger (genus *Mellivora* of the weasel family) works with the honey guide, *Indicator indicator,* a bird. The bird scouts out bees' nests, then leads the honey badger to them. The honey badger breaks up the nests, and badger and bird share the honey. (Walker et al. 1975).

Diagram the relationship between the honey badger and the honey guide and demonstrate the mutual dependence. Conjecture how this relationship could have evolved, and what sorts of events could destabilize it.

11. *Regulation of Population Size—Faginism*

The young of certain species—notably freshwater fish—are an important supply of food for the mature members of the same species. This relationship is called *Faginism,* after David Copperfield's mentor. It is not the same as a relationship that might be called *Swiftism,* after Swift's "Modest Proposal," for in that case, the nourishment obtained from eating

Irish babies could not form a substantial part of the diet of the entire population, since the babies themselves were nourished out of their mothers (parasitism?). In Faginism—at least in successful Faginism—the young must have access to food sources not available to their elders, and so serve as collectors. Fagin could not have made a living by having the boys rob his own coffers.

Like other such relationships, Faginism depends for its success on not being too successful. If all young are eaten, there would be no next generation and the species would disappear. Otherwise, it is a marvelous adaptive mechanism, since as the number of oldsters declines, the number of surviving young increases, and vice versa.

Diagram Faginism and show how it stabilizes a population. Look for social phenomena (such as alumni associations), operating on the same or a similar principle, and discuss how they prevent themselves from becoming too successful.

12. *Block Diagrams*

In certain situations, block diagrams can be manipulated to produce equivalent, but perhaps simpler, diagrams. In the area of control systems theory, confined to linear time-invariant systems, the algebra of block diagrams is well worked out. For instance, two parallel blocks labeled A and B can be reduced to a single block labeled $A + B$, and two serial blocks labeled A and B can be reduced to a single block labeled $A \times B$. Give examples in which these transformations might be useful, and examples where, because of the failure of linearity or time invariance, they will produce erroneous models of the system. (Reference: DiStefano, Stubberud, and Williams 1967)

12

Types of Regulatory Mechanisms

. . . one who learns why society is urging him into the straight and narrow will resist its pressure. One who sees clearly how he is controlled will thenceforth be emancipated. To betray the secrets of ascendency is to forearm the individual in his struggle with society. (Ross, in Nett 1968:409)

In an earlier volume (Weinberg 1975), we discussed the tendency of people to categorize. There are two kinds of people—those who dichotomize everything and those who don't. In this chapter and the next, we're going to ignore our previous warnings on the dangers of dichotomizing and do some ourselves, for the sake of bringing order to the vast diversity of regulatory mechanisms. This order is more *for* our brains, and *in* our brains, than "out there" in the world. Don't get so involved in these chapters that you forget that.

It's important to remind ourselves of *why* we sometimes categorize, or lump, the fine distinctions in the world into broader classifications. We do so because of the *Lump Law*, which says, in effect, "If you want to understand *anything*, you mustn't try to understand *everything*." In these chapters, we'll try to understand just a few general and important things about regulators.

Conditional and Unconditional Mechanisms

Most plant chemical defenses are passive: parts of the plant must be eaten before a repellent effect is felt. Animals, by contrast, cannot afford to allow parts of their bodies to be ingested before a predator realizes its mistake, and their chemical defenses are frequently forced to immediate attention through sprays, bites, or stings. North Americans are all too familiar with the odor of butyl mercaptan, the defense spray of the skunk that effects a posthumous revenge on passing motorists. (Whittaker and Fenny 1971:759)

Any need as profound as the need to regulate is bound to be met in myriad ways by different systems. By applying the Parallel Principle, we may be able to put a little order into this chaos, though we shall also have to make some sacrifices to the Lump Law. As a first example of the kinds of classifications we will make in this chapter, consider the broad classification implied by the above quotation, into "active" and "passive" regulatory mechanisms. Passive mechanisms are there whether used or not, but active mechanisms are called into action conditionally. The tortoise shell is passive, for it is produced and maintained long in advance of any current threat. Should a rock hurtle turtleward, the shell protects simply by being there, but the shell is there even if no rock ever arrives. For the hare, on the other hand, running is the reply to flying rocks—an active mechanism, since the hare is not always running.

Any system, of course, might employ both types of mechanisms, even against the same threat. The tortoise draws into her shell, or walks—however slowly—to find better eating. The hare's fur coat regulates against the cold in a purely passive way. Unlike people with fur coats, she does not shed her rabbit fur each time she enters the hutch. Still, though any system will have both active and passive mechanisms in its plait, we tend to classify by some overall impression of "main" regulatory mechanism. As Whittaker and Fenny suggest, plants seem to us to regulate largely through passive methods, while animals are perceived as active regulators. Though neither generalization is entirely correct, the dichotomy is central to our conception of the world of living things.

The passive mechanism always operates; the active mechanism only operates "if." In a way, then, the so-called passive mechanism is the more active of the two, for it never rests in the way an active mechanism can. For our purposes the terms "active" and "passive" might best be replaced by the terms "conditional" and "unconditional," according to the principal advantage and disadvantage of each method.

Because it is perpetually operating, the unconditional mechanism tends to be inefficient. For efficiency, threats must be rather frequent and not too diverse, so that they can all be met by the same generalized mechanism. If the threats from the environment are multidimensional and intermittent, it may be far more economical to have rather specific mechanisms called upon only when the need arises—to obtain greater protection for the same average expenditure. Furthermore, if the threats tend to be contradictory, as we know they may well be, the only feasible method may be to call on different mechanisms under different conditions.

Where the environmental demands *are* contradictory, then, we generally find separate active mechanisms, according to the Polarity Principle. But going even further with the Parallel Principle, we might expect to find that if the disturbances are contradictory in the environment, the system may be coordinated in such a way that the regulatory actions are also contradictory. For example, Sherrington (1906) enunciated this arrangement for physiology in terms of the principle of "reciprocal ennervation," by which the nerves are arranged to inhibit muscles from working against their antagonistic muscles.

Conditional mechanisms—especially if carefully coordinated—have potential advantages in economy, but they do have their disadvantages. In particular, because they are conditional, they can easily be fooled. The snapping turtle may pull into its shell as a harmless cloud passes overhead, yet it may fail to pull in as we glide up quietly in our canoe. In the first case it regulates when it should not; in the second case it fails to regulate when it should.

In testing scientific hypotheses we distinguish between Type I and Type II errors. By the Principle of Indifference, it really doesn't matter which is Type I and which is Type II, but generations of students have been drilled to be sure they know that:

    I.   Type I error is believing the hypothesis when it is wrong.
   II.   Type II error is disbelieving the hypothesis when it is right.
       (Or is it vice versa?)

Conditional regulatory mechanisms are subject to two analogous errors— (I) acting when wrong or (II) failing to act when right.

A Type I error means that the system is operating a mechanism unnessarily. At best, this error costs a system some of the efficiency advantage that conditional mechanisms hold over unconditional ones. At worst, unnecessary action will interfere with other regulatory mechanisms to the system's severe detriment. A mouse may be so busy avoiding imaginary eagles that it fails to notice the cat. In human beings, one limiting case of Type I regulatory errors is paranoia; another is the sort of miserliness which leads a person, lying on a mattress stuffed with stocks and money, to die of starvation.

Although a Type I error may lead to a failure to exercise a second regulatory mechanism, a Type II error means that the mechanism itself fails to act. This failure may be more perilous than a simple Type I loss of efficiency, though the Plait Principle prevents us from drawing general conclusions about the balance between the Types I and II.

There must, of course, always be a balance, otherwise the regulation would be either unconditional, or not regulation at all. We can avoid acting foolishly by never acting at all, and we can avoid missing an opportunity by answering every knock. The problem of life—and the problem of a conditional regulatory mechanism—is to do reasonably well in spite of all the uncertainties in the world. The problem is not, as Goethe thought, to choose between thought and action, but to choose *when* to think and *when* to act.

Error-Control

> To drink without thirst and to make love anytime, madame;
> that's all that distinguishes us from the other beasts.
>
> Beaumarchais

To drink without thirst is clearly a useful, even pleasant, idea, but carried to excess it becomes a Type I error. Thirst is a signal that one of our important variables, the water balance of our bodies, is getting out of range. By waiting to be thirsty—rather than worrying about the long drive across the desert or the bars closing in half an hour—we may experience discomfort or forego pleasure, but we can avoid making a Type I error. In other words, we postpone the regulatory act until we actually detect a change in the identifying variable, so we will not act when there is no real threat. After all, there may be a Howard Johnson's halfway across the desert, or the bar may be open late tonight.

Even when we wait for error signals, we may make a Type I error, for not every small change represents a threat. Nevertheless, underlying the strategy of control by error is the fundamental assumption of continuity, so that any large change in a variable will be preceded by a small change—even though not every small change will be followed by a large one. If there is insufficient continuity to give time for error-control, the only chance for regulating is to *anticipate*. Once the fox has it by the leg, it is too late for the hedgehog to start digging. Once our car has broken down in the desert, it is too late to start thinking about packing water. Once the bar has closed, we can forget about another drink until morning. And once we make war, it's too late to make love.

Regulation, we see, is a difficult business, treading the thin line between Type I and Type II errors. No wonder we find such a spectrum of mechanisms, running from pure error-control to pure anticipation. Perhaps it will help us to understand this spectrum if we look at the general problem a regulator faces, as seen by the regulator itself. For instance, we may wish to *design* a mechanism to improve the control a system has over one of its variables.

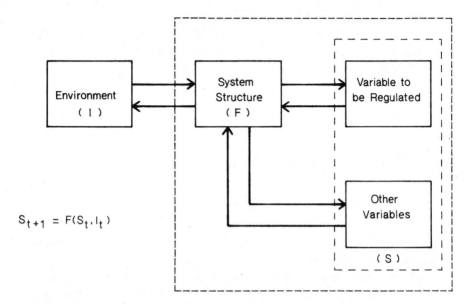

Figure 12.1. The general situation facing a regulatory mechanism charged with controlling the behavior of one variable in S, the state of the system.

The given problem looks like Figure 12.1. There is a system with a certain structure, which we assume for simplicity is cleanly separated from the system's state, which is a collection of variables. One of those variables is singled out for regulatory attention, perhaps because its behavior thus far is not meeting expectations, or just because the overall system performance would be more efficient if this variable could be better regulated. In human systems, political problems and business problems are often seen in this simple light—"just do this one thing a little better, leaving everything else alone."

We know that real systems problems are never that simple, but let's start from there and see where we are led. The diagram represents the equation,

$$S_{t+1} = F(S_t, I_t)$$

The new state (S_{t+1}) is determined by applying the system's program (F) to the old state (S_t) and the input from the environment (I_t). Our problem is that one of the components of S is not doing what we would like it to do, so we must change it in some way.

Figure 12.2 shows our problem in terms of creating a new system, an unconditional regulator, which will be "attached" to the other to control

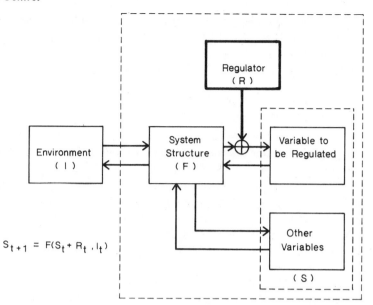

Figure 12.2. An unconditional regulator attempting to act directly on the variable to be regulated.

the errant variable. The regulator's output is shown going *directly* into the regulated variable, which corresponds to the equation:

$$S_{t+1} = F(S_t + R_t, I_t)$$

If, for instance, you're trying to raise the average income of poor families in a certain neighborhood, just *give* them each sufficient money to raise the average to the desired level.

Of course, only a fool could imagine the problem of regulating average income to be *that* simple. The only bigger fool would be the one who votes for the fool who's running for office with ideas like that. This, as Lincoln observed, includes all of the people some of the time. If only the politicians had to diagram their proposal, and the voters had then to draw the diagram of possible effects for that diagram, the world would soon be perfect.

Well, perhaps not perfect, but at least not surprised when after the money is given out, strange things begin to happen. Some people stop working, thus lowering the average again. Others get the courage to change jobs, and actually raise their income. Others take lower paying jobs that they *like* better, and sweatshop owners begin to complain. Loan companies go bankrupt; credit groceries raise their prices; real estate values see-saw as new people try to get into the neighborhood and cash in on the free cash. And, if the 101 direct changes weren't enough to spin everyone's head, the social workers, university researchers, and newspaper reporters move in to make *their* predatory living off the turmoil itself.

Perhaps it would have been better if we had acknowledged right from the beginning that the true picture was more like Figure 12.3, which shows that in attempting to change *one* variable, we risk changing every other variable, and even the environment. Although Figure 12.2 *implies* the effect that regulatory efforts may have on the rest of the system, the implication in Figure 12.3 is somewhat more direct. There, clearly interposed between the threats from the environment, are the various structures that transform what happens outside into what happens inside. There, too, the action of our proposed regulator is clearly shown to be just another "threat from the environment," or at least operationally equivalent to one.

In most systems of interest, the structure standing between inputs and effects has many facets. Some parts are directed explicitly toward regulating variables necessary for survival. Other parts may actually make regulation more difficult. Not everything a system can do is good for it.

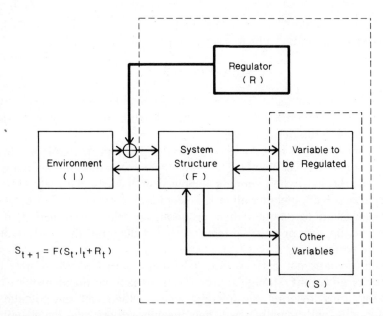

$$S_{t+1} = F(S_t, I_t + R_t)$$

Figure 12.3. An unconditional regulator acting indirectly on the variable to be regulated, showing how the system's structure has to be taken into account in deciding what to do to effect the needed change.

Some things are plainly bad, and some merely contradict other good things.

For instance, making love at any time may be terrific for the perpetuation of the species, but can be rough on the nervous system. Drinking without thirst may ward off depression and contribute to social solidarity, but puts a lot of strain on the liver. The body's regulatory mechanisms have to contend with these peccadillos, like it or not. They are all part of the structure that determines how the environment affects the interior.

In other words, the system's structure determines the *meaning of the environment* and will determine the meaning of any new attempt to regulate. The same environment will affect different systems differently. It is the system's structure that accounts for the difference, and which must be taken into consideration by any mechanism that intends to regulate the interior.

Within that structure, the regulator must take into account two parts. First, there is that part of the structure that converts outside happenings into inside reactions—the handsome tom turkey struts past, its tail feathers in full display, and the impressionable young hen's heart rate starts to rise. This relationship between what happens in the environment and what happens inside the system can be arbitrarily complex, and in something like a hen turkey, it essentially is. Yet the turkey's regulatory mechanisms seem able to cope with it, even though the turkey isn't very smart. How is it that although we're smarter than a turkey, and we don't fully understand how the turkey regulates its heartbeat, the turkey can actually do the regulatory job?

The turkey's trick is largely based on the simplification that *error-control* contributes. In Figure 12.4, we see a schematic diagram of how error-controlled regulation of the heartbeat could work. The heart rate regulator's first job is to *monitor* the heart rate. As long as it lies within acceptable bounds—no error—there is nothing for it to do. There could be an extravaganza of strutting toms surrounding our young hen, but if her heart rate doesn't waver, the regulator, like the hen, remains unimpressed.

In other words, the error-controlled regulator doesn't have to know what's happening in the environment. Nor does it have to know how various environments affect the heart rate. Therefore, the entire complexity of the environment and much of the complexity of the system itself need not be taken into consideration when deciding whether to take action.

This simplification eases the regulator's job considerably, but the rest of the problem is far from trivial. If, indeed, the regulator could act directly on the heart rate, without affecting anything else, then error-

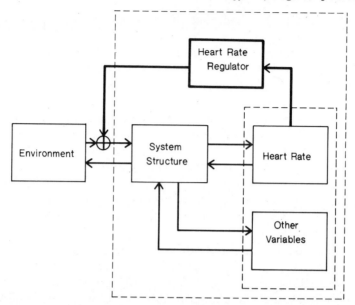

Figure 12.4. The error-controlled regulator works on the heart rate through the system's structure, but based on information taken from the heart rate itself.

control would be simplicity itself. But in most cases, part of the system's structure stands, as we have seen, between the regulatory mechanism itself and the variable it's trying to regulate.

The regulator must *somehow* take into account everything that stands between the detection of the increased heartbeat and the ultimate effect of bringing it back to normal. It's not sufficient merely to detect the *need* to slow the heart beat and say "Poof!" to magically get the needed reduction. Just the right combination of actions must be brought about, including some that might temporarily *raise* the heartbeat on the way to getting it lowered.

In order to take the correct actions, the regulator must "know" a great deal about the system's structure. This knowledge is translated into the repertoire of actions the mechanism will and will not initiate, the order in which they will be initiated, and the size of each. These actions might include nerve signals to increase respiration rate, the release of hormones into the blood stream, the activation of nerve impulses directed toward the neck muscles to cause the hen to look away from the exciting tom, or any of a number of other actions, possibly acting together.

Of course, in a system as complex and integrated as a living animal, the various operations may not be as cleanly separable as the neat boxes imply, but for the moment we may imagine that they are. In that case, the regulatory program might be conceptualized as a computer program, such as

1392. If heart rate is rising above acceptable level, then decrease amount of stimulating hormone released.
1393. If heart rate is falling below acceptable level, then increase amount of stimulating hormone released.

Notice that the program doesn't ask *why* the heart rate is rising or falling. It operates solely on the fact of an error in the regulated variable. Because of this, error control looks like a rather fool-proof scheme of regulation. Therefore, before we get too deeply entranced, we should apply the Count-to-Three Principle, which says, "If you can't think of three things that might go wrong with a system, you don't understand the system."

Limitations of Error-Control

> Look and see which way the wind blows
> before you commit yourself.
>
> Aesop
> *The Bat and the Weasels*

What limits the success of the error-controlled regulator? We can identify at least three important factors:

1. Delay in reaction to an error.
2. Interference from other mechanisms.
3. Loss of information on its own performance.

Let's take each of these in turn.

Any regulatory mechanism, not just error-controlled ones, will take a certain amount of *time* to affect the regulated variable. We call this time a *regulatory delay*. In Figure 12.4 by the time the released hormone reaches the heart muscle, the tom turkey may have vanished behind a bush. The heart rate may no longer be too high. It could actually be too low.

Figure 12.5 illustrates the sort of behavior that a regulatory delay can

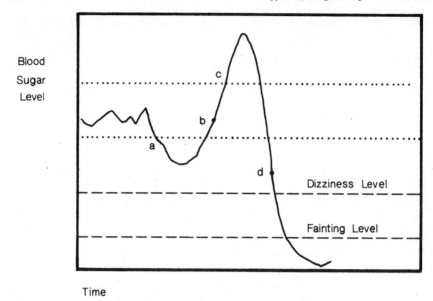

Figure 12.5. The effect of regulatory delay is to respond to excursion *a* at a time, *b*, when the response is in the wrong direction. The response to the excursion *c* comes at *d*, causing a wild deviation of the blood sugar from its desired range (dotted lines).

induce. The chronological graph traces the behavior of some regulated variable, such as, say, the level of sugar in Sally's blood. The dotted lines show the upper and lower limits allowed by the error-controlled regulators. Any excursion of the blood sugar outside of these bounds will trigger a mechanism to attempt to counteract the excursion.

Sally's blood sugar is, at first, fluctuating somewhat, but since it stays within the bounds, no regulation is triggered. At point *a*, however, her blood sugar falls below the acceptable level, and something happens. Perhaps that something is a vague message to Sally's brain to "eat something, maybe a candy bar." The message is delivered quite quickly, but it takes a few minutes to get to the cupboard, find the candy, unwrap a piece, chew it, swallow it, and digest enough of it to start raising the sugar in Sally's blood.

Finally, at point *b*, the sugar arrives, but alas, there have already been other factors at play to bring the blood sugar back into range. Perhaps some food she'd already eaten, but not digested, was on its way even as point *a* was reached. Perhaps her body was mobilizing sugar from its reserves. Perhaps Sally was in a hospital and receiving an intravenous injection of glucose solution. For whatever reason, the regulatory delay has produced a situation where the sugar arrives when it's no longer needed.

Of course, no harm may be done by this tardiness, since now the regulated variable is back under control. It could be, however, that the great rush of sugar coming from the candy is now sufficient to send the blood sugar *above the upper limit,* as we see at point *c.* This excursion will trigger other regulatory programs, dedicated to *reducing* the blood sugar, perhaps by accelerated production of a sugar-breaking enzyme.

Unfortunately, this enzyme production *also* requires a regulatory delay. By the time the enzyme has built up to an appreciable level, at point *d,* the candy bar's sugar is all gone, and the sugar level may actually be *below* the acceptable minimum. Yet here comes an enzyme designed to reduce the amount of sugar still further! Sally's blood sugar could well plummet, as shown, causing her to feel the symptoms of hypoglycemia—which just means "blood sugar too low." She may get dizzy, disoriented, or even faint. Such are the possible consequences of regulatory delay in an error-controlled regulator.

Can anything be done to counteract regulatory delay? Can we prevent it, at least, from taking a slight perturbation, as at *a,* and producing a runaway oscillation, as at *d* and beyond? One approach is to *reduce the sensitivity* of the regulator to *slight* excursions of the variable. If the dotted lines in Figure 12.5 were further apart, the excursion at point *a* would not trigger Sally's desire for candy. In this way, the entire syndrome might be nipped in the bud. What the regulator is doing, in this case, is sacrificing some of the quality of regulation of *small* variations, in return for some protection against *large* variations.

But can't the regulator *compensate* for allowing wider variations to trigger its action? Such compensation would be an application of the *Brain-Eye Law,* which says that "to a certain extent, mental power can compensate for observational weakness" (Weinberg 1975: 96). The "eye" here is the power to observe small variations. The "brain" is the power to take appropriate action to counteract these variations.

One way to compensate for the less sensitive observation would be to use a better *quality* mechanism for effecting the change in blood sugar level, one that used a better understanding of the structure between regulator and regulated variable. But if we had such a better quality mechanism, we would have to imagine it was used in the first place. Such a mechanism might have been available, but at a greater cost. In that case, the penalty for compensating for less sensitive variation would be *increased cost* of operating the regulatory system.

If we're not willing to pay the cost of a better quality mechanism, another way to make up for the later reaction is to respond *faster* with the same sort of mechanism we were previously using. In other cases, we may succeed with the same speed, but using *more* of whatever we are doing. A good example of this choice is the way you correct the steering of an automobile, when you discover you have drifted out of your lane. You can turn back quickly with a small turn of the wheel, or more slowly with a larger turn. After about the same interval, the total correction will be the same, but the later you start turning, the faster or more massive your turn has to be.

Fast or massive reactions can be costly to produce, just like more sophisticated mechanisms. Although this potential loss of efficiency is serious enough, a greater peril results from the nature of the system itself. A large change, or a fast change, becomes part of the internal environment that the *other* regulatory mechanisms have to contend with to do *their* job successfully. If a system is continually making sudden, massive, reactions to regulate a single variable, the problem for the other regulatory mechanisms will be severe.

This interference of mechanisms is why we so often find, in terms of the Plait Principle, a number of mechanisms "brought into action successively." Many smaller mechanisms are easier for the rest of the system to adjust to than a single massive mechanism would be. The interference effect also explains why, when there is a *fast* mechanism, it tends to be *small*, and when there is a *large* mechanism, it tends to be *slow*.

Interference with and from other mechanisms is thus the second limitation to the effectiveness of the error-controlled regulator, and, indeed, to the effectiveness of *any* regulator. Were it not for possible interference, each single regulatory mechanism would be able to make as large or as fast a response as it needed to regulate its variable in an optimal way. Instead, the system must strike a *regulatory compromise* among its own internal actions, not just among threats originating in the environment.

To illustrate how a system can be damaged by its own regulatory mechanisms, consider the way one human society defends itself against another. Such defense often takes the form of a series of barriers, one nested inside the other. These barriers may be physical—like walls or cannons—or psychological—like propaganda or laws—or both—like secret police or a standing army.

As each barrier is breached by the antagonist, a more severe one is raised to prevent further advances. If the invader is repelled only at a late state, the regulatory mechanisms themselves may have seriously changed the society. Many soldiers may have been killed. Stocks of precious materials may have been depleted. Citizens of "enemy" origin may have been put in prison camps. Civil liberties of the entire populace may have been removed, as by censoring the press, issuing identity cards, or requiring pledges of loyalty.

Paradoxically, in societies identifying themselves as "free," personal freedoms are usually the first variables to be sacrificed in the effort to regulate others. There always seems to be a minority, seething with fury that other people are allowed to say what's on their minds, ready to step forward at the flimsiest excuse to seal their lips. To the extent that free speech does, indeed, hamper a nation's defense, this minority serves as an important regulatory mechanism, ever ready to come to the aid of its country. To the extent that free speech is part of the *identity* of a "free" country, this censoring regulator is a lurking menace to survival. It all depends, as usual, on what the observer thinks is important to the system's identity.

Anticipation

> Where there is no vision,
> the people perish.
>
> Proverbs 29:18

Regulatory delay and interference between regulators are limitations that error-controlled regulators share with other forms of regulatory mechanism. The third limit to their success, however, is unique to error control, and can be understood merely by a logical analysis of the implications of error control itself. This limitation is due to the loss of information the regulator gets about its own performance as it gets better and better.

Imagine that we have an error-controlled regulator that has been growing increasingly successful. "Successful" means that the fluctuations in the regulated variable grow smaller and smaller. But these are the same fluctuations the error-controlled regulator uses to trigger its own action. If the regulator actually did a perfect job, there would be *no fluctuations whatsoever*. Therefore, there would be no information flowing along the arrow from, say, "heart rate" to "heart rate regulatory program" in Figure 12.4—no information from the regulated variable to the regulator.

Thus the *perfect* error-controlled regulator would sever its own source of operating information, as suggested in Figure 12.6. Without this feedback of information, the regulator would no longer be error controlled at all. Then where would it get its information needed to regulate? According to the diagram, it would have to be getting its information from the spirit world, in the form of magical visions of the future.

This limitation on error-controlled regulators means that every error-

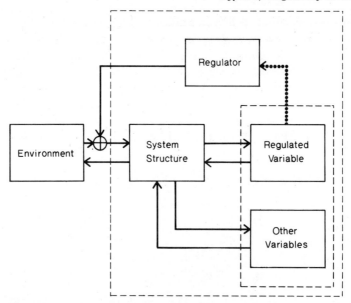

Figure 12.6. The "perfect" error-controlled regulator cannot actually be realized because perfect regulation would sever the flow of information from the identifying variable to the regulator.

controlled regulator has to *allow* a small amount of oscillation in the regulated variable. You can experience this oscillation as muscle tremor by performing the following experiment. Touch your finger to the bottom edge of a lampshade or other horizontal object. Keep your finger *just touching* the underside of the object, but *not pressing* on it. The sense of touch is the error signal. No touch means too light; pressure means too heavy. You will find it impossible to keep your finger exactly at the "no error" position. All voluntary muscles are controlled by a feedback loop ("reflex arc") regulating the opposition of two muscles, and thus exhibit such a tremor when trying to remain absolutely fixed.

Is this deficiency of the error-controlled regulator significant? This depends on the regulatory problem. If you try to keep your finger absolutely fixed *in space*, you probably won't notice the tremor, which was made obvious only by placing the finger just touching a solid object. In the same way, you probably never have noticed that your entire body sways from side to side as you stand "still." In certain kinds of precision work, this tremor or swaying may prove harmful, necessitating the use of tools to achieve the necessary performance. In normal activities, we have adjusted the activity to the oscillation, so there's no problem, unless we've taken alcohol or some other drug which changes the response characteristics of our feedback loop. Then we may noticeably sway from side to side, and our performance on the touch test will be obvious to all, except to us, since our sensory apparatus is likely to be similarly disturbed.

If error control were the only type of regulation, all regulators would be subject to this type of tremor, though it could often be made acceptable by careful design—at greater cost. Fortunately, there are other sources of information that the regulator can use, besides information on its own errors of regulation. To do a better job than error control, at least at eliminating oscillations, the regulator would have to *anticipate* what was going to happen to the regulated variable. If it *waits* for the error, as the error-controlled regulator does, then it will *have* to oscillate somewhat.

What the *anticipatory* regulator needs is information that comes *earlier in time* than an actual change in the regulated variable. If the system contains variables within the system whose behavior can be used to *predict* excursions in the essential variables, they can be used as the basis of regulation. By interposing a set of variables that do not *indicate* the loss of identity, but *anticipate* it, a system gives its regulatory mechanisms more time to act. Pain from a cut finger does not mean a person is dead, or even close to death. It does mean that a person *may* wind up dead if nothing is done.

The system of Figure 12.7 could represent the body's response to a cut finger. It could equally well represent the economic minister's response to a cut in the prime interest rate, an alcoholic's response to a cut in the liquor supply, or the military's response to an enemy missile cutting the protective radar screen. The hope is that before the threat from the environment has time to influence the identifying variables, the intervening

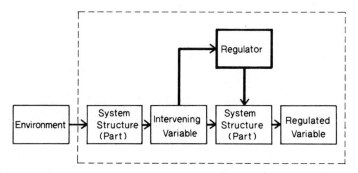

Figure 12.7. Anticipatory regulation based on information from variables that intervene (in time) between environment and the regulated variable.

variables will trigger the regulatory program to take action. The alcoholic orders another case of booze early enough to arrive before the last bottle is empty. The general orders the civilian population underground early enough to be buried alive, rather than roasted or crushed.

Systems of all sorts strive to create opportunities for anticipatory regulation. In economics, everyone is trying to develop or discover "leading indicators" to predict the future swings of the economy. In medicine, the Nobel Prize may await the researcher who discovers or creates a method of early detection of cancer. In "modern" warfare, great sums are lavished on "early warning" systems. Unfortunately, they're probably not early *enough*. Once the "early warning" signal has been triggered, the end of our identity is already winging on its way. For the military mind, this may qualify as anticipatory regulation, but to the diplomat, it represents the final collapse of all regulatory efforts.

Fortunately for those of us living under the threat of nuclear holocaust, anticipatory regulation need not be restricted to information from within the system itself. The diplomat need not sit on the shore and watch for missiles coming over the horizon but can sit in a conference overseas and watch for much earlier signs—in the environment itself. Figure 12.8, contrasted to Figure 12.7, shows that sensing the environment may provide much extra *time* for regulatory actions to take effect. Sensing a frown at a banquet gives many more life-saving response *options* than sensing a missile at your border.

Sailors predict the weather by looking at the color of the sky. Speculators predict the market by listening to the color of political speeches. If these informational inputs from the environment are sufficiently correlated with other inputs of importance, the regulatory mechanisms can use them to anticipate changes, as shown in Figure 12.8. The regulatory

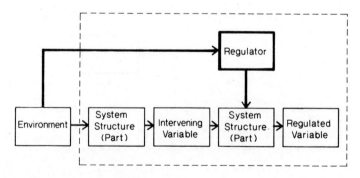

Figure 12.8. Anticipatory regulation based on information from the environment presaging a threat to the regulated variable.

strategy of *anticipation* can be pushed as far ahead in time as we can make reliable predictions about the relationship between events in the environment and events in the system. We shall next see what is required to make such predictions.

QUESTIONS FOR FURTHER RESEARCH

1. *Economics*

In attempting to regulate some economic variable, or a whole economy, it is necessary to sense the direction of movement of certain variables, such as prices, unemployment or employment levels, production, saving, or imports and exports. As Morgenstern (1955) points out, each such variable is subject to some uncertainty—of the order of 10 percent or more for typical measurements. Therefore, it is possible that if, say, unemployment for a given month is measured to increase by x from 4.0 to 4.3 percent, it could mean an annual increase of 3.6 percent, or it could actually be a decreasing unemployment marked by errors in measurement. Discuss how economic policies intended to regulate unemployment could be adjusted to overcome such measurement errors. Also, discuss the limits which each such policy is likely to encounter.

2. *Political Science*

In conducting a campaign, a candidate is trying to obtain a majority of the votes in her region. She is, however, plagued by certain problems in trying to influence the vote. For example, she may take surveys, but she is uncertain of their accuracy, and, even if the results are accurate, she is not sure which of her tactics is causing the swings in voter preferences. Also, her constituency consists of several types of voters, not all of whom will be swayed by the same considerations. What other problems does a candidate face in trying to regulate the voters' behavior, and what strategies are available to her? What limitations are there on these strategies, in the sense that there are general limits on all regulatory strategies?

3. *Political Science*

Discuss the structure of the United States government in the light of the principles of regulation discussed in this chapter. What are the variables being regulated? What are the formal mechanisms, as described in the Constitution, for regulating these variables? What extra-constitutional mechanisms now exist and participate in this regulation?

13

Regulation and Environment

Phenomena, then, are definite relations of bodies; we always conceive these relations as resulting from forces outside of matter, because we cannot absolutely localize them in a single body. For physicists, universal attraction is only an abstract idea; manifestation of this force requires the presence of two bodies; if only one body is present, we can no longer conceive of attraction. . . . In the same way, life results from contact of the organism with its environment; we can no more understand it through the organism alone than through the environment alone. It is therefore a similar abstraction, that is to say, a force which appears as if it were outside of matter. (Bernard 1957:75)

It's fairly easy for most people to accept the idea of sensing the environment in order to improve regulation. Most people have "an ear to the ground," "a finger in the wind," and "an eye on the horizon." But the role of the environment in regulation is not limited to a passive sending of omens and auguries to our sensors. Far from being a sterile monologue, the relationship between a system and its environment is rather a dialogue—if not always cooperative then at least communicative. Like all good practitioners of the fine art of conversation, the environment and the system take turns at sending and receiving. Both of these activities serve regulatory functions.

Acting on the Environment

It is difficult for a person familiar only with the relatively rugged growth of . . . microorganisms to appreciate the enormous sensitivity of mammalian cells when removed from their natural habitat. (Puck 1972)

Up until now, we have made two major classifications of regulatory mechanisms—conditional versus unconditional, and error-controlled versus anticipatory. In all our examples, we have more or less assumed that the system had to solve its problems by acting *on itself*. This assumption is unnecessarily restrictive and eliminates many of the most interesting regulatory mechanisms, as well as many of the most familiar. We therefore have a third great dichotomy—between mechanisms that *act on the system* itself and mechanisms that *act on the environment*. The first category we call *internal* regulators; the second, *external*.

When the level of profits is falling, a business may elect to regulate *internally* by cutting costs, or *externally* by increasing sales. When the level of your blood pressure is rising, you may elect to regulate *internally* by taking pills, or *externally* by avoiding stressful situations. Readers familiar with business know from experience that when profits fall, the first thought is usually to cut costs. Readers familiar with American medicine know from experience that when blood pressure rises the first thought is usually to take pills. Both instances are examples of our mental tendency to "localize in a single body" everything pertaining to a system.

Designers of regulated systems exhibit the same bias toward doing something *inside* the system, rather than letting the system do something to the environment. Researchers likewise tend to look *inside* their system for the source of regulatory behavior. Even as we see this tendency in others, we slip into it ourselves, as when Puck speaks of "the relatively rugged growth" of microorganisms, as if the range of environments in which they could survive were any less narrow than that of mammalian cells. True, there are many more environments that will support microorganisms, but when compared with all possible environments we could imagine, their number is paltry.

The problem is, we don't imagine very many environments, but take for granted the normal range of conditions found on the extremely well regulated surface of this particular planet. For instance, 99.99999 . . . percent of all *imaginable* environments would have no water, but we tend not to imagine those, until we're in one. Nobody who hasn't lived in a place where all water had to be carried in or manufactured can appreciate what crimes are committed when Americans brush their teeth.

Once we stop taking the environment for granted, we begin to see how *common* external regulation is. Systems employ external regulation independent of the source of regulatory information. We commonly see both error-controlled (Figure 13.1) and anticipatory (Figure 13.2) external regulation.

For instance, a child crying for food is acting on its environment—the surrounding adults who will respond to the crying by bringing food or coming to see what's wrong. What triggers the crying? It could be hunger signals from inside the body (Figure 13.1) or the setting sun indicating it's nearly time to eat (Figure 13.2).

Suppose we expand the system to include both child and parents. The

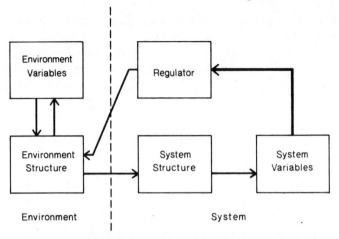

Figure 13.1. Error-controlled regulation through the environment results when the regulator acts on the environment based upon information received from the system's variables.

parent is acting on the family's environment—the interlocking set of tribal obligations—when borrowing food from a kin. What triggers the borrowing? It could be the child crying for food (Figure 13.1), or the knowledge that relatives are coming, bringing more mouths to feed (Figure 13.2).

Suppose we expand the system even further, to include the entire tribe. The tribe is acting on the environment—the entire geographic region in

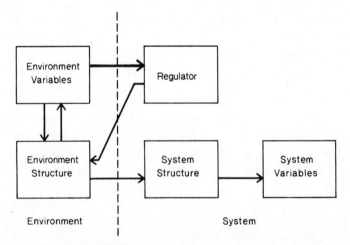

Figure 13.2. Anticipatory regulation through the environment results when the regulator acts on the environment based upon information received from the environment.

which the tribe moves throughout the year—when it migrates. What triggers the migration? It could be the empty larders because grazing has been poor (Figure 13.1) or the first signs of cold weather and the need to move the herds to the lowlands (Figure 13.2).

In each of these examples, we see the possibility of *part of the environment* to one system being *part of the system* to another. We know this must happen because of the Principle of Indifference: calling it a "system" or "environment" won't change what's actually going on. Yet by the Principle of Difference, we know that looking at the system-environment split in a different way can lead to quite different conclusions about "reality."

Take the case of the "early warning system" which we previously concluded was actually a "too late warning system." If we shift our view of *who is being warned,* we might obtain a different understanding of how this system can regulate the temperature of international hostility. Suppose we consider the "warning" of the early warning system to be a warning to our *enemies*—the environment—rather than a warning to *us*—the system.

From this perspective, new insights emerge. By *building* the system, we are acting on the environment. We are sending a *message* to our enemies: "If you act upon us, we shall act upon you." This action is *very* early, much earlier than the counter-strike itself. This action is also very *small,* giving the environment more chance to respond rationally. But *their* view of what is system and what is environment will help determine the meaning of "rationally."

Quite likely, their generals will interpret our building of the early warning system as a signal from *their* environment. They may decide to act upon their environment (us) before their environment (us) can act back (Figure 13.2). Generals on both sides are incessantly urging that the anticipation be pushed ahead of the "error" signal—the actual launching of enemy missiles. They want a system that acts *on* the environment on the basis of anticipatory signals *from* the environment.

What signals? Perhaps the announcement of a new offensive weapon. Or a new defensive weapon. Or being called the "running dogs of capitalism." Or having their ambassador seated at a low prestige position at the conference table. Wars have started with less anticipation.

The Environment Regulation Laws

Our conflict is not likely to cease so soon as every good man would wish. The measure of iniquity is not yet filled . . . Speculation, production, engrossing, forestalling . . . affording too many melancholy proofs of the decay of public virtue . . . and too glaring instances of its being the interest and desire of too many who would wish to be thought friends, to prolong the war.

George Washington
Letter to a Friend, March 31, 1779

The Principle of Indifference can be applied to the system-environment relationship in another way. We can reverse the role of system and environment, to make us aware that the environment is also a system and may exert regulatory influences on the system in attempting to maintain its *own* identity. Thus we find *mutually regulating systems,* a simple case of which is diagrammed in Figure 13.3.

Notice how the survival of the system and of the environment are tied up in one big feedback loop. This loop can lead to mutual destruction, mutual survival, or destruction of one or the other of the participants. It all depends on the actual relations between the regulatory mechanisms. In the case of the international missile systems, the system and environment are more or less symmetrical. In human systems, where one group forms the major environment of another, it's a morally sound position to argue from symmetry, as in "do unto others as you would have others do unto you." Moral questions aside, however, it's safer to assume that the only thing system and environment share is a desire to survive, by some means or another.

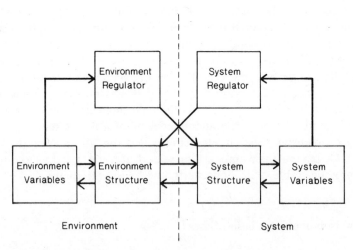

Figure 13.3. Because the environment is also a system, three can be complete symmetry in the mutual regulation of system and environment.

Generally, in real-life cases of mutual regulation, we find a ravelled skein of feedback loops, often leading to conclusions that seem to contradict the predictions of simpleminded analyses.

To take a relatively simple example of this complexity, consider the relationship between a person's environment and health, particularly concerning infectious diseases. People who live in small, isolated places are less likely to contract infectious diseases than people who live in densely populated cities. On the other hand, some diseases, like measles, confer lifetime immunity when contracted. If such a disease is more severe when contracted by adults, the city person is more likely to *have* the disease, but less likely to *suffer* from it very greatly. Perhaps the strategy for rearing children should be to rear them in the country, with monthly visits to the city throughout childhood. Hopefully, this mixed strategy would use both environments to best advantage for the child's survival.

Whenever a system regulates by acting on the environment, we won't be surprised if the environment changes as a result of the system's action. Conversely, if we notice that the environment changes as a result of the system's action, we have a clue about what sort of regulatory mechanism to look for in the system. For instance, in recent years, extremely obese people have begun to use radical surgical procedures for rapid weight loss. One such operation is to bypass most of the patient's intestine, thus drastically lowering the absorption of food without changing eating habits.

To the surprise of the physicians, a large percentage of such patients experienced striking changes in their most intimate relationships as soon as the success of the operation became evident. Spouses expressed extreme jealousy and felt utterly inadequate sexually with respect to their "new" mate. The physicians were surprised because their model of obesity failed to take the spouse—a major portion of the obese person's environment—into account.

The reaction of the spouses in these circumstances reveals the presence of previously unrecognized regulatory mechanisms acting through the spouse to keep fatty's weight nice and high, as shown in Figure 13.4. Previously, medical thinking centered exclusively on the other regulatory loop—fatty's own regulatory activities—which naturally led to conclusions about the necessity for radical surgery. With this new insight, the physicians' job is more complex, but at least they know where to look. Perhaps they can avoid unnecessary intestinal bypasses on fatties by substituting more expensive lobotomies on the insecure spouses.

The obesity situation is an instance of a general heuristic for discovering regulatory mechanisms that sustain a "problem" in the face of concerted efforts to rid ourselves of it. First imagine that the problem were

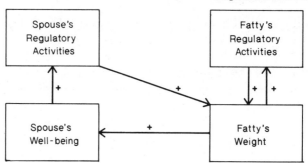

Figure 13.4. Fatty's weight turns out not to be maintained solely by Fatty's own regulatory mechanisms. Something the spouse has been doing has had a regulatory effect on keeping Fatty's weight up.

instantly removed, as we might reduce obesity by actually slicing out layers of subcutaneous fat. Then imagine what would happen in response to this sudden change. Anything that will be changed is a candidate for a regulatory mechanism that keeps the problem going.

For instance, if all the smokers in the United States suddenly gave up the "filthy habit," it would directly affect a number of industries and services:

1. The entire tobacco industry.
2. Trucking (no tobacco shipments).
3. Farmers (one less cash crop, more competition).
4. Advertising (loss of tobacco ads).
5. Medicine (fewer patients).
6. Welfare (more people living to old age).
7. All government agencies (loss of tax revenue).
8. Fire departments (fewer fires).
9. Fire insurance (lower premiums, thus lower profits).
10. Life insurance (longer life for insured customers).
11. Health insurance (better record initially, but lower premiums eventually).
12. Candy manufacturers (increased sales).
13. Ashtray manufacturers (decreased sales, followed by retooling to make candy dishes).
14. Printers of antismoking literature.

You can undoubtedly add a few of your favorites to the list.

Each of these industries, whatever their sentiments, has a stake, positive or negative, in the continuation of tobacco use. By the Parallel Principle, we can infer that *any* system that regulates through its environment is likely to have some portion of its regulatory strategy devoted to maintaining that environment, whether consciously or not. This heuristic is so important we'll pose it in the form of a law, the *First Environmental Regulation Law:*

Systems that live by the environment, live for the environment.

In some cases, the connection is obvious. If the system consumes some environmental ingredient, that ingredient is likely to come into short supply—leading to starvation—if the system doesn't act to help the environment preserve itself. We can easily imagine how domestication of animals might have arisen from the observance of taboos on hunting females and infants—even though females and infants may be easier prey.

Conversely, if the environment is used as a repository for some waste product, it may become clogged, leading to poisonous "foodback." A company selling magazines door to door requires an environment relatively free of "distrust." If its policies and practices include any kind of shady dealing, we can imagine that it is, in effect, excreting distrust into the environment, just as we excrete hundreds of "disposables" into our water. Because its own excrement of distrust will poison the market it feeds on, it must adopt a strategy that regulates distrust, lest its business be poisoned.

If that strategy is to move quickly from town to town, you can easily draw your own conclusion about whether to order their magazines. Otherwise, they will have to obey the First Environmental Regulation Law and devote themselves to maintaining goodwill in the community. They will join the Better Business Bureau, purchase goodwill advertising, contribute to the fire fighters' circus for crippled children, sponsor a slow-pitch softball team, and make pious statements to the press about fly-by-night out-of-town companies that sell magazines and never deliver.

In such instances, it's obvious that the system lives *by* the environment, but there is another class of systems that sometimes attains even greater intimacy with its environment. Many systems in human societies are specifically and conspicuously dedicated to *protecting* other systems *from the environment*. The *Second Environmental Regulation Law* says:

Systems that live against the environment, live by the environment.

The observation is simple enough. Without narcotics peddlers, there would be no narcotics agents; without illness, there would be no doctors;

without enemies, there would be no military. Though the *expressed* reason for the existence of the military is protection against the enemy, the generals, like parasites, will extinguish themselves if they are *too* successful. Advocates of disarmament who fail to reckon with this regulatory force are bound to underestimate the difficulty of achieving peace. But then, without war, there would be no antiwar movement.

The Regulatory Model

I know that when I get into my car there are some things I must do to start it; some things I must do to back out of the parking lot; some things I must do to drive home. I know that if I jump off a high place I will probably hurt myself. I know that there are some things that would probably not be good for me to eat or to drink. I know certain precautions that are advisable to take to maintain good health. I know that if I lean too far backward in my chair as I sit here at my desk, I will probably fall over. I live, in other words, in a world of reasonably stable relationships, a world of "ifs" and "thens," of "if I do this, then that will happen . . ."

What I have been talking about is knowledge. Knowledge, perhaps, is not a good word for this. Perhaps one would rather say my *Image* of the world. Knowledge has an implication of validity, of truth. What I am talking about is what I believe to be true; my subjective knowledge. It is this Image that largely governs my behavior. (Boulding 1956:5–6)

If the thermostat is the cliché of regulatory models, error controllers are the clichés of thermostats. We can enliven this weary example, and in passing improve upon the regulator, by putting a sensing device *outside* the house to measure the *outdoor* temperature.

With an outdoor sensor, the thermostatic mechanism can *anticipate* indoor temperature changes and counteract them before they even begin. But such a regulator must not only *sense* outdoor changes, it must *predict* how these outdoor changes will affect the temperature inside. The thermostatic regulator shown in Figure 13.5 must, in short, know the properties of the system structure.

This "system structure" consists of all the factors that affect the way outside temperature is converted into inside temperature, such things as the material and construction of the walls, the number and placement of the windows, the frequency of opening and closing doors, the insulation in the roof and walls, the color and condition of the paint, the tightness of the joints, and the habits of the occupants.

Take just one factor, insulation. For a poorly insulated house, the in-

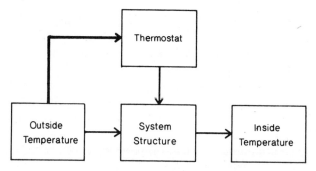

Figure 13.5. Anticipatory regulation can be illustrated by a thermostat that senses *outdoor* temperature.

terior temperature will quickly follow the exterior, and the regulator would have to start up the heating soon after the outside temperature dropped. With better insulation, the response must be either slower or smaller, lest the house overheat in anticipation.

It's clear that the amount of information the regulator needs about the house could be enormous. In Boulding's terms, "it is this Image that largely governs" the behavior of the regulator. Right or wrong, true or false, accurate or inaccurate, complete or incomplete, the anticipatory regulator must have some sort of image, or *model*, of how its various actions will affect the regulated variable.

The rewards of anticipatory regulation depend on the quality of this model, a term we prefer to "image" when we don't want to imply any *consciousness* on the part of the regulator. If the model is poor, the behavior may be worse. What counts is not the refinement or sophistication of the model, but its correspondence with the actual facts of the matter. If, for instance, the thermostat predicts a drop in internal temperature, and thus starts heating, it will waste energy and create discomfort if the drop does not actually come, or is less severe than the model predicted.

Acting when not needed is a Type I error that may exhaust the system's resources. Failing to heat when the temperature is freezing may be more economical, but could also prove fatal to the occupants of the house. Or, if they're smart, this Type II error could prove fatal to the thermostat system with the outside sensor and the defective model.

Systems that commit fatal errors, even systems that habitually waste resources, don't last long. We don't find many of them alive and kicking, which explains why the Parallel Principle is such a powerful heuristic. With long-surviving systems, we can argue from environmental and system structures to regulatory mechanisms, or vice versa. For instance, if we find a ubiquitous regulatory mechanism, we can be confident that the environment for which it was "designed" will also be widely distributed. There are few air-conditioners in Alaska, and few furnaces in Florida.

Each part of the regulatory mechanism must contribute its share to the parallelism with the environment. If the regulator anticipates by sensing in the environment, then the sensor must be attuned to significant signals from the environment, signals that carry reliable information on which to base a model.

Eyes typify a ubiquitous mechanism permitting all sorts of anticipatory regulation. Animals lacking vision (or some equivalent anticipatory system, like the sonar of bats) are limited in the speed with which they may move. Because they must use *contact* with an object to signal its presence, the blind would destroy themselves in collisions if they moved with any great speed. Congenitally blind people who gain sight in adulthood experience difficulty grasping the idea that the source of a light stimulus is not *touching* their eyes (Von Senden 1960).

Why does vision work? Why can it almost completely replace touch as a source of information from the environment? We can perhaps best understand by considering some cases in which vision is handicapped or useless. In a fog, we slow down. When we swim in muddy waters, we lose our sense of direction and may drown by going deeper when we think we are ascending. Fog and mud particles block the passage of light through air and water, which are normally transparent to the wavelengths we see.

Isn't it a marvelous coincidence that air and water allow visible light to pass? After all, not all wavelengths penetrate both, and some penetrate neither. Or consider another kind of failure of vision. The invisible man doesn't bump into things, but other people bump into him! How lucky we are that there aren't many transparent solid objects like the invisible man.

Well, it's not really a matter of luck. If *most* solid objects, even if *many* solid objects, were as transparent as the invisible man, our eyes wouldn't be nearly as valuable as sources of reliable information for the model we use in regulating our movements. Conversely, if the most common gas and liquid media through which we move—air and water—were not usually transparent to certain wavelengths, then eyes that sense those wavelengths would be hardly any use at all to a regulator of movement.

Superman is supposed to have X-ray eyes that enable him to see through most solid objects. Fortunately for Superman, he has super powers that protect him when he bumps into things, or is it his normal vision that supplements the X-ray eyes as regulatory sensors? Otherwise, his ancestors wouldn't have survived as long as they did on the planet Krypton.

We don't have any comic book heroes whose ancestors had eyes that could see through stone, but not through air and water. If we did, you can be sure that they, too, would have to have super powers to have survived with such a useless mechanism. Or perhaps instead of "faster than a speeding bullet," they were "slower than a crawling blindworm."

These environmental qualities which have come to be paralleled in the regulatory mechanism are often conceptually invisible to us as well. The transparency of water to visible light isn't nearly as perfect as that of air, so the ancients had a pretty clear idea about the existence and nature of water. But air was a puzzle until very recently. It was difficult to think of air as a "thing" or "substance." Wind, perhaps, but not air.

Only when one of those constant environmental qualities changes do we notice—with a bang. Most of the time, what we can't see can't hurt us. But if you put a pane of glass in the middle of a field, birds will fly into it and kill or injure themselves. James Thurber knew that, but many modern architects learned the hard way. So did a few people who have walked right through large glass doors.

Human organizations—bureaucracies—are especially adept at continuing to regulate on the basis of environmental qualities that have long since disappeared. In 1957, a study was made of data processing procedures of a large military base. Though World War II had then been over for a dozen years, at least 20 percent of their effort was devoted to producing reports that had been inaugurated as *emergency* measures during the war. Nobody had bothered to stop them. The war was over, but its memory lingered on, just as the memory of a long-gone vegetarian environment lingers on in our bellies in the form of an appendix.

The Game of Regulation

A Man and a Satyr became friends, and determined to live together. All went well for a while, until one day in wintertime the Satyr saw the Man blowing on his hands. "Why do you do that?" he asked.

"To warm my hands," said the Man.

That same day, when they sat down to supper together, they each had a steaming hot bowl of porridge, and the Man raised his bowl to his mouth and blew on it. "Why do you do that?" asked the Satyr.

"To cool my porridge," said the Man.

The Satyr got up from the table. "Good-bye," said he. "I'm going: I can't be friends with a man who blows hot and cold with the same breath."

Aesop
The Man and the Satyr

The simplest error-controlled regulator doesn't take its information from the environment, but from the regulated variable itself. Consequently, it needs no particular model of the environment to predict what's about to happen to the regulated variable. It does need something of a model, though, in order to regulate at all. Its sensory model says that the regulated variable is at least somewhat *continuous*—that the immediate future will resemble the immediate past.

Without *some* continuity, *no* regulator could regulate. Only in lotteries do we find utter lack of continuity, for we cannot win tomorrow's lottery on the basis that number 666 came up today. That's exactly the feature we *want* in a lottery, because we don't want *anyone* to be able to regulate it. And to regulate, we have to know that the situation we're trying to control hasn't changed *completely* by the time we've taken action on the basis of present information. If there's even a *little* bias in the lottery, someone may find out and cheat—that is, regulate.

That's the *minimum* model—a little continuity between present and future. In order to *act* effectively on the basis of this continuity, the regulator also has to know something of the structure that stands between *its actions* and the *reactions* of the regulated variable. Thus the regulator must have a model of at least part of the system itself.

Of course, if the regulator acts *through the environment,* its model must have even more information—information about how its actions are related to the actions of the environment. A frog has to be able to distinguish between the signs of an owl about to swoop down on it and a fly perched above its nose. But it must *also* know which one of them it should try to eat. In other words, it must recognize the "moves" of the environment, and it must also know a few "moves" of its own. Especially it must know which of its moves to make in response to each of the environment's moves. The frog must recognize the *states* of the environment and the *states* a frog can exhibit, but it must also have a *structure* that relates the two in a way advantageous to itself.

The *complete* model, either explicitly or implicitly, must contain the *relationship* between what has happened and what is to be done. The tsar may order the troops to fire into the crowd of workers when a conciliatory gesture would have sent them humbly back to their factories. On the other hand, conciliation may simply fan the flames of revolution. In the long run, the tsar's model must not deviate too far from the underlying "reality," otherwise he soon ceases to be the tsar. Indeed, there may be *no* action the tsar can take to remain enthroned—no regulator can work all the time.

To design a regulator, a good starting point is to think of this "game" the regulator plays with the environment. We write down the environment's moves and the system's moves, then try to relate one to the other in such a way that the system "wins" by keeping the desired variable within range. Note well that the system "winning" doesn't necessarily mean the environment "losing," except in the special case where the environment is out to destroy the system. In some regulation problems, the environment is benefitted by the system "winning." In most, the environment is indifferent to the system and its problems.

A convenient form for describing this "game" is the "payoff matrix" of game theory (Williams 1954). Each *column* of the matrix is labeled according to the possibilities for the behavior of the variable *sensed*. In an error-controlled temperature-regulating system, we might have three columns corresponding to Goldilocks' three categories—too hot, too cold, just right.

Each *row* would be labeled with the various *moves* the regulator has at its disposal, such as, furnace on, cooler on, both off. At the intersection of each row and column, in the matrix itself, we can write the cost, or "payoff," of taking each action under each sensed condition.

Figure 13.6 shows such a payoff matrix from which we might begin the design of an error-controlled temperature-regulating system. The payoffs do not need to be exact, at this level of design, but can simply be given qualitatively, as in the figure.

The matrix can be thought of as describing a game the error-controlled regulator is playing against the variable being regulated. The variable makes a "move"—that is, it selects one of the columns. Since this regulator is error-controlled, it selects *its* move *after* the variable has moved, by selecting one of the rows. The "outcome" of the game, for that pair of moves, is the entry in the chosen column and row—the "payoff." For instance, in Figure 13.6, if the variable "chooses" the "just right" column and the regulator "chooses" the "cooler on" row, the outcome is "accept-

Regulator Action \ Inside Temperature	Too Hot	Just Right	Too Cold
Furnace On	Very Bad	Acceptable	Good
Furnace Off Cooler Off	Bad	Good	Bad
Cooler On	Good	Acceptable	Very Bad

Figure 13.6. The payoff matrix for an error-controlled thermostatic regulator which can cool or heat.

able." This outcome corresponds to running the cooler when it may not be necessary—a Type I error which may be wasteful, but not outright fatal.

Those acquainted with set theory or the preceding volume (Weinberg 1975) will recognize the game matrix as the *Cartesian Product* of the two sets of possible behaviors. All this means, to the uninitiated, is that *every possible pair of moves* is represented in the matrix by one and only one entry. Therefore, if all the entries are filled in, the entire structure, or rules, of the game is represented in the payoff matrix.

The payoff matrix does not describe the *playing* of the game, but only the *consequences* of the various plays. How the game actually comes out is determined by the structure of the regulator—which moves it chooses under which sensed circumstances. This part of the regulator's structure could be characterized by a *program* that gives its "strategy," for example:

1. If "too cold," then turn on cooler.
2. If "too hot," then turn on furnace.
3. If "just right," then turn on furnace.

This particular program characterizes an especially *poor* regulator, for each choice will produce a "very bad" or, at best, "acceptable" payoff. A slightly better program would be:

1. If "too cold," then turn furnace off and cooler off.
2. If "too hot," then turn furnace off and cooler off.
3. If "just right," then turn furnace off and cooler off.

Though this "regulator" hardly deserves the name, at least it's better and cheaper than the previous one. It doesn't add heat to an already hot room, nor does it cool an already cold room. And it does give us a minimal standard of comparison against which we can measure other proposed programs. If they can't do better than nothing at all, then they ought to do nothing at all.

In fact, investigators often fail to give proper recognition to regulatory mechanisms simply because they are unconditional and therefore may not be seen "in action" under the investigated circumstances. But laziness is a regulatory strategy, too, a fact which we should perhaps emblazon on our minds with the *Lazy Law:*

When regulating, doing nothing is doing something.

Indeed, in a salubrious climate, this second regulator might provide the ideal solution to the problem of temperature regulation. Since regulation is a game, we cannot say how good a regulator is unless we also know something about the environment's program. Therefore, the "quality" of a regulator is a relative concept.

The payoff matrix, however, shows the *bounds* on the performance of the regulator program, given the various possible actions in its repetoire. Suppose we were trying to build a regulator using a simpler thermometric device, one that could only detect two ranges—"too hot" or "too cold." The payoff matrix for this system (Figure 13.6) now has only two columns, since there are only two detectable temperature states. With this limitation, what can we say for this program?

1. If "too hot" then turn on cooler.
2. If "too cold" then turn on furnace.

The best we can say with assurance is that it would be "acceptable," for sometimes it would be heating or cooling unnecessarily.

Regulators built under the restrictions of Figure 13.7 can't do as well as a good program built with a "just right" sensor, though of course a bad program could do worse. The reason for the difference is not hard to see in the matrix of Figure 13.7, for since the sensor can only detect two states, one of the three strategies is bound to be superfluous. Indeed, with such a restricted sensor, we need not waste our money on providing three possible actions, for the payoff matrix in Figure 13.8 would permit just as good a regulatory program as that of Figure 13.7.

The study of regulatory models may be carried out at no less than three levels. First, we can simply ask the static question, "How does it work?"—a question we can answer by examining the program. Second, we can ask the more dynamic question, "How could it work *better*, with the same components?" For this question, we need the game matrix in

Regulator Action	Too Hot	Too Cold
Furnace On	Very Bad	Acceptable
Furnace Off Cooler Off	Bad	Bad
Cooler On	Acceptable	Very Bad

Figure 13.7. The payoff matrix for a thermostatic regulator with a limited sensory ability, and thus unable to sense "just right" temperatures.

Inside Temperature / Regulator Action	Too Hot	Too Cold
Furnace On	Very Bad	Acceptable
Cooler On	Acceptable	Very Bad

Figure 13.8. The payoff matrix for a thermostatic regulator with limited action ability, so either furnace or cooler must be in operation at any given time.

some form or another, showing both the possibilities and the payoffs—the value to the system of each possibility. Third, we can ask the even more dynamic question, "How could it work better given *different* components?"

The second question—the underlying question of game theory—asks "How can we change the *program* part of the model?" The third question—the underlying question of the study of adaptive systems—asks "How can we change the underlying *game matrix?*" Thus, we find the study of regulatory models turning back to our three fundamental systems questions: "Why do I see what I see? Why do things stay the same? Why do things change?"

QUESTIONS FOR FURTHER RESEARCH

1. *Animal Domestication*

The tangled web of relationship between system and environment is well demonstrated in Konrad Lorenz's (1964: 1–19) hypothetical account of how dogs came to be domesticated. He suggests that a symbiotic bond was formed between early human groups and packs of golden jackals (*Canis aureus*)—food scraps for the jackals, and an "early warning" defense system for the people. Using some of the ideas in this chapter, speculate further on this domestication process. Fill in the details and bring the story up to present times.

2. *Revolutionary Journalism*

Lenin once made the following argument against numerous local newspapers in Russia trying to carry on revolutionary work. Although the local papers by their very numbers would seem less likely to be wiped out by the authorities (the aggregate strategy), they are in fact less experienced at the arts of concealment than a single central paper run by a small devoted staff of professional revolutionaries (the specialist strategy). Of course, if the single paper does get wiped out by the authorities, the entire movement suffers a fatal blow.

Develop these ideas into a theory of revolutionary newspaper strategy, taking into account the different environments ("free press," "censored press," etc.) in which a revolution might be brewing. (Reference: Lenin 1929)

3. Feedback Versus Nonfeedback Systems

An open-loop control system is one in which the control action is independent of the output.

A closed-loop control system is one in which the control action is somehow dependent on the output.

In order to classify a control system as open-loop or closed-loop, the components of the system must be clearly distinguished from components that interact with, but are not part of the system. For example, a human operator may or may not be a component of a system. (DiStefano, Stubberud, and Williams 1967)

This distinction is evidently strongly influenced by the point of view of the observer. Control systems theorists speak of the calibration of open-loop systems, by which they mean the preprogramming that must take care of reacting to the input to produce the appropriate output. Discuss the observations that preprogramming is:

1. Feedback on a different time scale, in many cases.
2. Present in all feedback devices, in all but the "feedback" mechanism, and even there, at a meta-level.

4. Gambling by Computer

A simple example of the detection of constraint in the environment is the game of penny matching.

If the two players do not flip the coins, but choose each play explicitly and simultaneously, then that one will win who can most successfully model the constraint in the opponent's play. Show how a computer might be programmed to play this form of penny matching, and how well one computer strategy would do against another. If possible, program your strategy, and test it against human beings, to see if it can model constraint in them better than they can model constraint in it.

Note for those who have never matched pennies: The payoff matrix for the game looks like this:

My Play	Your Play	
	Heads	Tails
Heads	+1¢	−1¢
Tails	−1¢	+1¢

5. *Anthropology*

An important characteristic of culture is its transparency to the society that lives by it. Consider the analogy that the system of meanings and rules that is "culture" is like a perfectly clean sheet of glass. Think about the ramifications of this model, both for the members of that society and for the anthropologist who would understand their culture.

6. *Mutual Regulation*

In giving Figure 13.3, we violated the Law of Happy Particularities (Weinberg 1975:42) and gave no examples of system-environment pairs exhibiting this particular pattern of mutual regulation. Develop examples of systems with this structure drawn from

1. International politics.
2. Social psychology—marriage, for example.
3. Biology—such as the interaction of two species.
4. Education—between teacher and pupil.

14

When the Model Fails

It is thus rather a puzzle how far, if at all, what is rather confusedly called control by error is relevant to the working of [regulatory mechanisms] in the systems of men or societies. We have to distinguish carefully between two possible meanings of the phrase. In so far as we can compare what is happening with what ought to be happening . . . , we are kept informed of what is going wrong and perhaps the rate and direction of what is going wrong. Control by error in this sense is valuable and not uncommon. What we are not told with any certainty, if at all, except in the simplest and most often repeated cases, is what contributions our past actions have made to this result. Control by error in this sense is much more valuable—and much more rare. (Vickers 1968:464)

Because the anticipatory regulator has an extra step, or extra model, it has much more potential versatility than the error-controlled regulator. But the price of this versatility is the additional possibility of going astray with an incorrect or incomplete model. In this chapter, we shall explore some of the ways models fail, and what regulators do to protect against those failures.

The Fundamental Regulator Paradox

Most of the luxuries and many of the so-called comforts of life are not only not indispensable, but positive hindrances to the elevation of mankind.

Henry David Thoreau
Walden

A seed lying dormant in the ground must know whether the increased moisture in the soil is a sign to increase or decrease its metabolic activity. The seed wants to know if spring has come. In some environments, wct soil announces spring. In others, the dampness may be from autumn rains. If the seed crosses signals, it will sprout in autumn or fail to sprout in spring.

If the seed could somehow regulate by internal error control, it could never make such mistakes, lacking as it would the extra model of the environment. For the same reason, however, it couldn't anticipate the effect of spring.

We've already seen why the feedback regulator can never be absolutely perfected, since it would thereby sever its own supply of information. Is the anticipatory regulator similarly limited?

At first thought, this limitation seems not to apply. The anticipatory regulator uses its *model* to stay ahead of the changes in the regulated variable. Of course, if there is any flaw in the model, the regulator's effectiveness will be thereby circumscribed. Thus the question is, "Can the regulator eliminate all flaws in its model?"

If a model is to be improved, it must be on the basis of information from the past, not by magic. The model may be adjusted within the lifetime of the system, or the system may have inherited a model from successful ancestors. Information of either sort can come only from experience at regulating—it is nothing more nor less than error control on a slower time scale. Thus the same paradox applies to possible perfection of a model as applies to possible perfect regulation from *any* error-controlled regulator. The more effective it becomes, the less information it receives about the performance of its model—information that might permit improvement of the model for future regulation. This, then, is the *Fundamental Regulator Paradox:*

> *The task of a regulator is to eliminate variation, but this variation is the ultimate source of information about the quality of its work. Therefore, the better the job a regulator does, the less information it gets about how to improve.*

Put in more memorable words, the Fundamental Regulator Paradox says:

> *Better regulation today risks worse regulation tomorrow.*

Not only does this paradox prevent regulators from being perfect, but also the better they get, the more difficult it is for them to improve.

The Fundamental Regulator Paradox carries an ominous message for any system that gets too comfortable with its surroundings. It suggests, for instance, that a society that wants to survive for a long time had better consider giving up some of the maximum comfort it can achieve in return for some chance of failure or discomfort.

This lesson is easiest to see in terms of an experience common to anyone who has ever driven on an icy road. The driver is trying to keep the car from skidding. To know how much steering is required, she must have some inkling of the road's slickness. But if she succeeds in completely preventing skids, she has no idea how slippery the road really is.

Good drivers, experienced on icy roads, will intentionally *test* the steering from time to time by "jiggling" to *cause* a small amount of skidding. By this technique, they intentionally sacrifice the perfect regulation they know they cannot attain in any case. In return, they receive information that will enable them to do a more reliable, though less perfect, job.

Pilots use a similar technique when they are about to land. They joggle the controls to see if reduced air speed has killed their effectiveness. If the plane fails to respond, the pilot increases speed to regain sufficient control for a proper landing.

Without the jiggling or joggling, the driver or pilot doesn't have any way of knowing if success comes from skill, or from the momentarily benign environmental conditions. The jiggling and joggling tactic is an *investment in model improvement.* Not only does it refine the model of the transitory environment, but also, even more important, it refines the model of the system's own regulatory ability, what Vickers calls the "contributions our past actions have made to this result."

The second refinement—of the system's model of *itself*—is "the more valuable," and much more rare. This isn't surprising when we realize, as Vickers goes on to say, that:

The two senses are, in fact, points on a continuous scale. Control by error enables us to compare the way things are going with the way we want them to go; and in doing so, it tells us something about the effect of our previous action. But this something may be anything from nearly everything to nearly nothing. (1968:464)

The "continuous scale" Vickers speaks of is a time scale. Regulation by error control might be called regulation by *recent error*. Anticipatory regulation could then be called regulation by *remembrance of errors past*. Both are applications of *The Axiom of Experience* (Weinberg 1975: 141): "We only know the future by the past." And, indeed, since any regulator has a model, there is always the choice to make between believing the past (the model) and believing the present (the error, or lack of error).

Which to believe, past or present? The question is as old as the division between parents and children, between conservatives and liberals. What is our experience trying to tell us when something goes wrong? As Vickers implies, we'll never know for sure. Indeed, we could have derived that insight from the *Principle of Indeterminability* (Weinberg 1975: 214): "We cannot with certainty attribute observed constraint either to system or environment."

We're all amused to hear of the farmer who, in 1968, said he would vote for Nixon because he had voted Republican in the previous two elections and the weather had been pretty good for eight years. But we seldom know better than the farmer about the success of our *own* regulatory efforts. We may know they succeed, but what makes us so confident that the success comes from us? When the patient recovers, it was brilliant surgery; but when the patient dies, it was an Act of God.

Noise

Those interested in communication have been aware from the very beginning that communication circuits or channels are imperfect. In telephony and radio, we hear the desired signal against a background of noise, which may be strong or faint and which may vary in quality from the crackling of static to a steady hiss. In TV, the picture is overlaid faintly or strongly with an ever-changing granular "snow." In teletypewriter transmission, the received character may occasionally differ from that transmitted. (Pierce 1961:145–146)

Those interested in communication share the conceit of surgeons, farmers, and the rest of us. When their communication circuits or channels do not behave the way they expect them to, they say, "The circuit or channel is imperfect." They do not usually say, "My model of the circuit or channel is imperfect," but that would be an equally valid conclusion. It might also be a more *productive* conclusion.

Consider the matter of "noise" on phonograph records. Many of us with beloved record collections are somewhat fanatic about keeping the records free of dust, scratches, or anything else that produces noise. Many companies, some quite large, make their profits from this obsession with noise on phonograph records. You can buy special cloths, brushes, cleaning fluids, and no end of thingamajigs to attach to your record player to feed your obsession. But the noise remains, albeit subdued.

A few years ago, someone had the idea that the noise was not on the record, but in *our model* of records. Actually, a great deal was known about the nature of the sounds produced by dust and scratches on records. Engineers also knew how to *subtract* a known electrical signal from an unknown one. They didn't have to know the exact form of the subtracted signal but only some of its general characteristics.

The general characteristics of sounds produced by scratches are quite different from the general characteristics of sounds produced by musical instruments or voices—that's why those sounds annoy us so much. And, when both sounds are picked up from the record, they are converted by the stylus and cartridge into electrical, rather than sound, signals. There-fore, we already knew how to get rid of noise, as soon as we lost our il-lusion that the noise was "on the records." With that stumbling block out of the way, a "black box" was developed and put on the market to elimi-nate scratch noise from records.

The device functions roughly as shown in Figure 14.1. An interesting sidelight, illustrating the Principle of Indifference, is the switch on the black box that allows its owner to choose to eliminate *either* the noise or

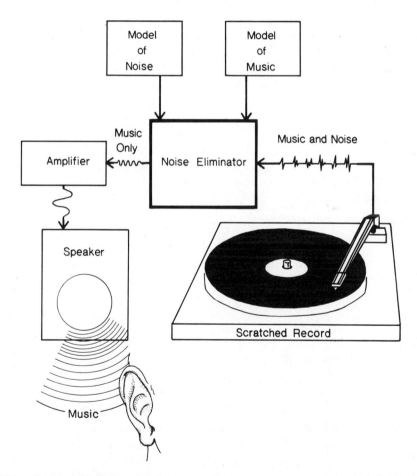

Figure 14.1. A black box added to a record playing system to remove scratch noises.

the music. Thus, we can listen to scratches on a record, without being bothered by the music.

A sound system can be viewed as a regulator whose job it is to keep the sound emissions in the range of what we call "music." We now see that the failure of a sound system to deliver noiseless music can be considered a failure of its model of its environment, a record. The name systems thinkers give to imperfections in models is, appropriately, "noise."

Noise, in the systems sense, is that part of the *structure of a system,* as reflected in the behavior of that system, which is *not* part of the observer's model. Thus, for instance, if our demographer's model predicts the three components of a population will be

$$(280, 380, 820)$$

but it actually is observed to be

$$(267, 378, 841)$$

then the noise is

$$(-13, -2, 21)$$

Note that it doesn't really matter *why* a model fails, but only that it *does* fail. The demographer may feel confident that the theory *could* incorporate the

$$(-13, -2, 21)$$

by, say, expanding the structure matrix to 15 by 15. But—whether for lack of patience, lack of intelligence, or lack of money—the present model does *not* account for the

$$(-13, -2, 21)$$

so it *is* noise. Quite frequently, we stop short of eliminating all possible discrepancies in a model. The noise eliminator of Figure 14.1 could have been built years earlier, but the *cost* of the components was then too high for the market to bear. The plunging cost of microcomputers changed this device from a theoretical possibility into a marketable commodity.

Every regulator faces the same problem concerning its choice of models. Recall our discussion of primary and secondary simulations, where a primary simulation was always a more *refined* model than a secondary simulation but sometimes had to be dismissed because of excessive costs. We should carefully note, however, that a coarser model need not have more noise than a finer one. For instance, the demographer—if looking only at total population—might have predicted 1500 merely by assuming rough continuity from one period to the next. The error in *this* model's prediction would be 20, less than the error in the third *component* in the three-element model. On a percentage basis, it is *much* less. Indeed, after a certain level of refinement, a model is likely to have more difficulty with noise.

Some systems thinkers are fond of accounting for this difficulty by speaking of some kind of "intrinsic" noise. By this they seem to mean some "real" unpredictability that is "out there" in the world, in the sense that *no* model, no matter how refined, could *ever* account for it. And, indeed, there may be such "intrinsic" noise in the world, but by the Principle of Indeterminability, we know we can't *prove* that it's intrinsic. To be sure, there are physical phenomena on an atomic level that quantum theory tells us we can *never* predict. But where did quantum theory—another model—come from? It came, like all models, from past information, so it, too, can have noise, in which case it *might* be wrong about intrinsic noise. Probably not, but it's good to learn not to *worship* anybody's model.

If we want to get rid of noise in quantum theory, or any other model, we have to spend money or energy. Indeed, we have billions of dollars, marks, rubles, and other currencies invested in the model we call "quantum theory." Why do we pay so much for these models, if they can't be perfect anyway?

Quantum theory, like any widely applicable scientific theory, can be used in building "low-noise" models for many different regulators. By eliminating noise in these regulators, we can possibly obtain a desired degree of regulation much more efficiently—or possibly obtain regulation we couldn't have at all with less precise models. Conversely, if our model is poor, we can often compensate by paying more for operating the regulator. In other words, we can make up for ignorance with energy, if we are sufficiently clever.

Noise in Communication Systems

Theorem 11: Let a discrete channel have the capacity C and a discrete source the entropy per second H. If H is less than or equal to C there exists a coding system such that the output of the source can be transmitted over the channel with an arbitrarily small frequency of errors . . . (Shannon 1949)

In 1949, Shannon's mathematical theory of communication appeared in the scientific world blushing with the promise of a high-school valedictorian. Inasmuch as we previously remarked that this promise has in many ways remained unfulfilled, we'd now like to lay some credit where credit is due. Information theory does not deal with the full problem of regulation, but with a subproblem of great importance to any regulator. The subject of information theory is the "communication system," a device that is illustrated schematically in Figure 14.2.

Roughly, the idea of a communication system is just what our intuitive understanding says it is—to get information from the source to the destination. For our present purposes, the communication system is the simplest possible regulator. It tries to do nothing to the information from the environment but *reproduce* it inside the system (the destination). Because this is the simplest regulatory job we can imagine, it's a good place to start thinking about the effects of noise on regulation, and how those effects can be overcome.

Even in more complex regulators, whenever there is a need to get information from one part of the regulator to another, there will have to be some sort of communication channel. It might be a nerve pathway in a body. It might be a telephone link in a corporation. Whatever form it takes, any noise added to messages passing over it will certainly compound the regulator's general problem with noise.

Shannon showed that under very general conditions, messages could be sent over channels with arbitrarily great protection against noise. His Theorem 11 showed that, contrary to intuition, the cost in channel capacity to do this was arbitrarily small—if we could only design the correct scheme. Of course, Shannon did not consider the *cost of designing the scheme* as part of the cost in channel capacity. This cost corresponds roughly to the cost a system invests in building its regulatory models.

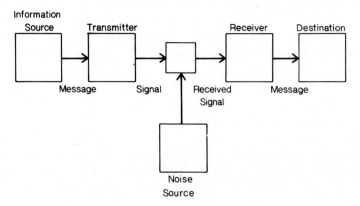

Figure 14.2.   Schematic diagram of a general communication system.

Whether it will be worth finding such a scheme for a given channel will depend on what we're using the channel for.

Information theory does *not* tell, in most cases, even today, *how* to overcome noise in this optimal way. Since 1949, many brilliant schemes have been developed for combating noise in communication systems. When we have control of the information source as well as the destination, we can add special codes to the messages, a subject which goes under the name of "error-correcting codes" (Lin 1970).

The more general case, however, faced by a regulator, is when information has to be taken from an environment that has not been so kind as to add error-correcting information designed for our convenience. The extraction of information under these conditions is studied under the name, "signal detection theory" (Swets 1973). These subjects are well worth studying if you have the mathematical background, but we still await the definitive tutorial work for nonmathematicians. For our purposes, it will serve merely to convince you that it is *possible* to recover information that has been "lost" to noise, if we are willing to pay. How to make the recovery optimal is best left to another book.

If we are attempting to send a telegraph message from Aix to Ghent, and if we are aware that the message may be corrupted by noise on the way, we can achieve some measure of protection by sending the same message several times. If the noise is not too systematic, we shall begin to see the original message emerging by keeping our eyes on what stays relatively fixed in the different received versions.

Thus, for instance, if we receive:

> WHAT HAWX GOV WQOUGXT.
> WYAT HOTH BOG WROUGXX.
> WHAX HATH GAD BROUGHT.

we might begin to suspect that the message being sent was

> WHAT HATH GOD WROUGHT.

This doesn't seem very mysterious, but we must notice that the technique only works because we do know *something* about what is taking place between Aix and Ghent. At least we think we do.

What do we think we know? Among other things, we assume that there is nobody plugged into our communication system who is *deliberately* trying to mislead us. If that were the case, the actual message going in at the source could be *anything*—perhaps

> SEND UP TWO CORNED BEEF ON RYE.

Indeed, if there was malice in the channel, we might see

WHAT HAS BOB BOUGHT?
WHAT HAS BOB BOUGHT?
WHAT HAS BOB BOUGHT?

and believe *that* was the original message. The "intelligence" community has developed the art of malicious message sending to an exquisite degree, so that they would make the above false messages even *more* convincing by adding a little noise for the sake of realism:

WHAX HRS BBB BXUGRT?
XHAT AAS BOB BOQGHR?
WHOT HAZ VOB POURHT?

By the time we have invested ourselves in deciphering this into

WHAT HAS BOB BOUGHT?

we are unlikely to suspect that the original message was something else entirely.

Unless there is maliciousness somewhere along the way, repetition, though slow, is sure. In natural systems, the assumption of a nonmalicious environment is generally good, but there are plenty of examples in which one system tries to fool another by imitating a third. Just watch a tabby cat trying to make the sparrow think he's actually just some grass. Or watch the zebra make the lion think she's just some shadows among the trees.

The tabby and the zebra aren't as good at this sort of thing as the National Security Agency, so the sparrow and the lion have a chance to detect their presence by looking carefully enough, perhaps from slightly different angles. The situation is captured quite nicely in the simulation of a *very* noisy telegraph transmission shown in Figure 14.3. The left column contains the sequence of messages. They've been distorted in transmission in such a way that each letter has only one chance in five of being transmitted correctly in any one repetition. This corruption is serious enough to destroy our ability to spot the message at a glance, as we did previously. Can the message still be extracted?

The right column shows the receiver's guess based on all the information received up to that point. In each character position, the receiver places the letter *most frequently seen* in that position so far. At first, there is very little to go on, and the guess is no more clear than the messages. But as the messages accumulate, a distinct pattern begins to emerge. Finally, the full message is seen, though the system may not quite *recognize* that it has the full message. It may continue to change its guess, as we see in the next-to-last line.

Message as Received   Best Guess So Far

```
CWA VMSMKOKGE MHTI        CWA VMSMKOKGE MHTI
VTVPVQMOXKWMH DOJF        CTAPVMMMKKKGE DHJF
OFNQ OCGJBXDRPNYGQ        CFAPVMCGJBKDE DHGF
AQPKKPCSX M SMMOXE        AFAKVMCGXBKDE MOGE
DFRTEHSXYOHLEKVTIS        AFAKVHCGXOHDE MOGE
WAESNMC DZ FACPOAX        AFAKVMCGXOHDE MOAE
WHEV OUFK IYGWTOCV        WFEKVMCEKOHDE MOAE
IUREMESAPOJJIRBCNG        WFEEVMCAKOHDE MUAE
RAAI ESBTFGGO DPYE        WAAE ESAKOGGE DOAE
GHKU LMETZ QEWFOXN        WAAE ESEKO GE DOXE
 GYCUBS NU YWNDKYO        WAAC ESEKO GE DOXE
HGJMSWO VMGFKJSOY         WAAC ES KO FE DOYE
LVQTOPMLZPHUODJHLO        WAAT ES KO FE DOYE
MCDL IL PDNCEKEOY         WAAT ES KO FE DOYE
HTNXAIS PLOPEFWINE        HAAT ES PO FE DOYE
ICSFEFXCGNXQ UWUP         HAAT ES PO FE DOYE
VSOPYHACSU VCZVHHY        HAAP ES PO FE DOYE
WHUMFYZ THUULRWVD         WHAM ES PO FE DOYE
LRAZKWGXZXUBFCCKYE        WHAM ES PO FE DOYE
LTLQ AD IMGTSXFVNT        LHAM ES PO FE DOYE
FWFHWQMTLCLGROCBMJ        LHAM ES PO GE DOYE
XHST EOTTDVKEBNOTX        LHAT ES TO GE DOYE
.SHAP PARBPFUHKJOSE       LHAP ES TO GE DOYE
PSMEAVSP DMOEOPYNM        LHAP ES TD GE DOYE
SSGTWIWLBRFSJDVOTO        LHAT ES TD GE DOYE
TTAZRNSWVMAXMFJLWY        LHAT ES TD GE DOYE
WAATWAQOAPRTSUAPNQ        WHAT ES TD GE DONE
XKVRBIZXMXBVBSRHAE        WHAT IS TD GE DONE
RZVREGANTOXXFGWOEY        WHAT IS TO GE DONE
HJSBGPJ YUVBVIIBFN        WHAT IS TO GE DONE
OHATRFEZKOUXIQVLIN        WHAT IS TO GE VONE
YMLZ EN K KBCCVKDC        WHAT ES TO BE VONE
DTPTPIBKG DRRJDOEE        WHAT IS TO BE VONE
ZQLKRSMVGRNAV ZGVP        WHAT IS TO BE VONE
VNRKFPSHDBBVHGNICD        WHAT IS TO BE VONE
WYIRFCBFWOJFSUAOTB        WHAT IS TO BE VONE
WBZT IE TORZULDJJT        WHAT IS TO BE DONE
WAALRQM UURYBMAQQN        WHAT IS TO BE DONE
WPAM EJQPBULI NEZE        WHAT IS TOUBE DONE
YA TBYO HC RP NSLV        WHAT IS TO BE DONE
```

Figure 14.3. Extracting a message from a very noisy teletype transmission by guessing that the most frequent character in a column is the actual character transmitted.

What we have here, in effect, is a practical demonstration of the Eye-Brain Law. The computer can't "see" the message very clearly, but eventually can figure it out. It takes time, and it takes energy, but it works. Of course, it may take too *much* time, and the cat may have nabbed the sparrow while it was still trying to figure out whether it really was a cat.

Repetition, as a general strategy, is very commonly used by regulating systems for extracting conditioning information from the environment. For example, an owl will not swoop toward the first mouse-scratch it hears, but listens to a short series of scratches in order to distinguish the mouse from any transient source of sound. A bear does not go into hibernation on the first cool day in the fall, nor does it emerge on the first warm day of late winter. In all such cases, possible noise is overcome by trading speed for certainty. The system responds more slowly in order to respond more surely.

The proper balance between speed and assurance is established by the relative cost of Type I and Type II errors. Waiting for many repetitions is most plausible when Type I errors—acting wrongly—are more costly than Type II—failing to act or acting too late. The predator, if it fails to pounce before the victim bolts, simply waits patiently for another victim who is not so fidgety. But if it attacks too quickly, before being sure of the nature and location of the prey, it simply frightens away one potential meal after another.

If the prey commits a Type I error, there is usually no second chance. Consequently, it will be to the sparrow's advantage not to wait to be absolutely sure it's a cat, not the wind, rustling the grass. This balance of error costs is so ubiquitous in nature that people have always characterized the hunter by steadiness and the hunted by skittishness.

The repetition need not be of exactly the same message, over and over. We often extract the same information from different sources, even different media. Thus the sparrow both hears and sees the cat, which greatly increases her chances of timely detection.

Whatever the form of repetition, the net effect is to raise the amount of energy put into the signals relative to the amount of noise. The signal is presumably always there, while the noise comes and goes. As we add more and more messages together, much like in Figure 14.3, the combined weight of the repeated message *eventually* drowns out the lesser weight of noise. The same computer program used in Figure 14.3 can extract a message from noise in which the true character appears only 1 percent more often than any other character in its place. It merely takes longer.

When the signal emerges from the environment, repetition is the *only* way to raise the "signal-to-noise ratio." If the system itself creates the signals, however, noise can be overcome without as much waiting. A bat sends out signals that will be reflected back to its ears by nearby moths. By making this signal more powerful, the bat can raise the energy in the moth's reflection, thus overcoming noise.

Unfortunately for the bat, this strategy also makes it easier for the moth to hear that there's a bat in the vicinity. Over countless generations, moths with bigger ears have frustrated the bats' efforts to raise the signal-to-noise ratio in this simple manner. The bat and the moth regulate one another, but that's the Regulatory Compromise the bat must accept.

Noise in Models

The facts, deliver'd by antient authors, are either so uncertain or so imperfect as to afford us nothing decisive in this matter. How indeed cou'd it be otherwise? The very facts, which we must oppose to them, in computing the greatness of modern states, are far from being either certain or compleat . . . 'Tis to be remark'd, that all kinds of numbers are uncertain in antient manuscripts, and have been subject to much greater corruptions than any other part of the text; and that for an obvious reason. Any alteration, in other places, commonly affects the sense or grammar, and is more readily perceiv'd by the reader and transcriber. (Hume 1752:211–212.)

It is not only in the immediate receipt of conditioning information that noise interferes with regulation. From past information the system builds its regulatory models. Because noise gets into the model, each action, even if called into play at the right time, may be wrong.

Consider, for instance, the action of a company trying to maintain or increase the sales of its product. One strategy is to modify the price, but in order to decide how to do this, the company must have a model of the collective buying behavior of its clients. The common economic model assumes that in order to increase sales, prices must be lowered, though products differ in their "price-sensitivity." In certain fields, however, over a certain range of prices, a decrease in price is accompanied by a *decrease* in sales. For instance, a lipstick which doesn't sell very well at one dollar may sell even worse at fifty cents, because for a product like lipstick, it is *prestige* which sells. Since all lipsticks are pretty much the same, prestige is judged by such externals as promotion, packaging, and price. Thus the lipstick manufacturer who attempts to regulate falling sales by price cutting soon goes out of business, the victim of a reversed model of his environment.

Some systems thinkers would be hesitant to call the error from such a defective model "noise," for the assumption is often implicitly made that noise is a *small* residue of unaccountable behavior. This definition, however, is based on observation of the external world, not on general principles. Naturally we find that for *most* systems the noise *is* a relatively small part of the regulatory models, because those systems with highly defective models do not last very long and are not likely to be seen. But when we build a new system, we cannot assume that it will automatically have accurate models just because other, older, systems do.

As we know, the strategy for overcoming noise is to expend energy. When complimented on the precision of the guidance systems in American rockets, one of the old German rocket scientists pointed out that the American systems had to be precise, since they had no energy to spare. The Russians, on the other hand, had much bigger rockets. If they made a mistake in navigation they had lots of thrust remaining to correct it. Of course, without the press of international politics, systems that take more energy to do the same job of regulation are at a disadvantage in the competition for survival.

Still, there may be situations in which the best available strategy is simply to overpower the noise, rather than to try to account for it. If we are trying to hire workers, it may not be worthwhile, or possible, to figure out the exact minimum wage they will accept. If we think their minimum is $6.50 per hour, we can offer them $7.50 and be pretty sure that they will accept, even if our model of their desires is a bit defective. If we are trying to catch a fish whose maximum strength is thought to be 50 pounds, we can be sure it won't break the line if we use a 100-pound filament. Such overpowering, of course, is not "sporting," which is why anglers try to take the maximum fish on the minimum line. They are not, contrary to some belief, trying to save money on equipment. If that were the case they would buy their fish. The fish in the market are caught by people whose living depends on *not* giving the fish a sporting chance.

In some situations, the method of overpowering noise may not be so simple. For example, consider the problem of adjusting the flame on a gas stove that has a sticky valve. We try to raise the flame, but the valve will not move until we apply so much pressure that it abruptly slips past the intended point. Trying to move it back, we have the same trouble, for we cannot predict the friction with sufficient precision. What we can do, however, is to use one hand to push *against* the direction we are trying to turn the valve. If we push hard enough, the frictional resistance becomes a relatively small part of the force we must overcome to position the knob. By working "against" our own goal, we make precise adjustment possible in the face of an unknown, but small, amount of stickiness.

This strategy, by which the system overcomes noise through supplying opposing actions, is essentially what is involved in the "factors having an opposing effect" in Cannon's Polarity Principle. If a system relies on an uncertain environment to supply the opposing factor to one of its regulatory mechanisms, that mechanism must have a much more refined model. By supplying its own opposing factor, it can get away with a much simpler model of the environment.

Fudging the Model

What this means in practice is that our response should always be to an uncertain image. We should not put all our eggs in the basket that happens to be around at the moment. In the jargon of the economist, our image of history should never be so free from uncertainty that we feel we can afford to dispense with liquid assets. (Boulding 1956:131)

Supplying a polar factor is but one special case of a more general strategy of ensuring against noise. As Boulding expresses it so nicely, this is the strategy of the Plait Principle—of using several baskets with a few eggs in each.

Boulding's "liquid assets" metaphor is particularly apt. The investor in possession of a noiseless model of the investment environment would have no need of cash reserves. Indeed, the *presence of reserves* in any system suggests to the perceptive observer the existence and magnitude of imperfection in that system's model of its environment or of itself.

Reserves, the contents of "buffers," represent the system's fudging its imperfect model by falling back on the most primitive of regulatory strategies—aggregation. If a business keeps a large cash reserve, the various parts of the business can spend as needed without constant reference to the level of cash. In effect, the reserve permits other regulators to operate within a wider range, even though the model is rather cloudy. Their job of regulation is thereby simplified, and the reduced expense of regulation may compensate for the extra expense of keeping the reserve.

Reserves do cost. Money in the cash reserve could be earning money if invested in productive assets. An extra large gas tank means fewer stops and less attention to the gauge but costs more fuel to transport the extra gas itself. In the end, there is a maximum amount of fuel a vehicle can carry, because transporting any more fuel over the total distance would burn more fuel than the extra fuel transported. A few well-placed kilos of body fat not only make a person attractive to some members of the opposite sex, but eliminate the need for eating with excessive regularity. But the fat reserve costs in several ways—extra work for body organs such as the heart, more energy to move the entire body a given distance, less mobility if the reserve gets beyond a certain point, and loss of attractiveness when the reserve becomes *too* conspicuous.

Some buffers are conspicuous, whether filled or not. In large medical complexes, we see huge waiting rooms designed to hold sufficient patients to smooth over the inability to predict how long each procedure will take. Patients complain about having to sit and read ancient (or even "antient") magazines, but they would also complain about the higher fees the center would have to charge if technicians and equipment were idle each time a procedure was finished early.

The same kind of queuing of customers is seen wherever the customers' time is assigned low value with respect to the time of the server. Doctors are very valuable—the AMA has carefully arranged it that way—and patients have nothing better to do anyway. Professors' time is surely worth infinitely more than the time of any group of students waiting to find out what happened to their term papers. Although some professor may profess to be anti–elitist, you need only examine the queues of students to verify the actual value system in force.

Actually, the professor and the doctor could avoid the queues if they shaped their environment better by sticking to a rigid schedule of appointments. Then, of course, they would be accused of "impersonal" or "cold" behavior, when all that was involved was a concern over giving the patient or student the most efficient service.

The underlying problem is not so much attitudes as it is noisy models. The medical center's inability to predict doctor or patient behavior will lead to waste which will come out in one form or another. One way is to increase the reservoir of patients for each doctor, but another would be to increase the reservoir of doctors. Since the doctors don't care for the second method, and since they control medical education, we patients simply have to put up with being part of the reserve.

It may not console the impatient patients, but their plight is the subject of a branch of mathematical study called queueing theory [for a brief introduction, see Coats and Parkin (1977)]. The objective of queueing theory is to permit the medical center to minimize the queueing under the fixed constraints of the situation, much as game theory permits us to study optimal regulatory programs, given the fixed constraints of the payoff matrix. Game theory does not allow us to change the rules of the game, and queueing theory does not allow us to increase the number of doctors, lower their pay, or invent new medical practices. Nevertheless, it does allow us to get the most out of a rather miserable situation.

Not all reserves are as conspicuous as huge rolls of fat or hundreds of patients angrily tearing magazines to shreds. When they are hidden from us, they can make the understanding of regulatory mechanisms rather baffling. For example, our atmosphere seems to be one of the best regulated systems around, but even today scientists are puzzled over what keeps the main active components within their narrow bounds. Take carbon monoxide, a fatal compound that would be most annoying to breathe in moderate concentrations. Human activities annually exude about 200,000,000 metric tons of carbon monoxide, but none of it seems to be concentrating in the atmosphere—yet. Where does it go?

One suggestion is that carbon monoxide is metabolized by a vast population of micro-organisms in the soil, a buffer with an estimated capacity of several times the current rate of production. But because nobody is sure that this buffer is the principal source of regulation, nobody knows whether we're at the edge of its capacity and heading for disaster.

One would imagine that our advanced civilization would know more about what regulates such a deadly poison. But carbon monoxide seems to be well regulated—so far—so few investigators have been eager to study it. When people begin falling dead in the streets—when the buffer capacity is exceeded—surely we'll have plenty of time to learn about the mechanism, won't we? With a sufficiently big buffer, no system needs a particularly refined model.

Discarding the Worn-out Model

The Human Motives: Survival and Security:
Pertaining to the Body:
Avoiding of hunger, thirst, oxygen lack, excess heat and cold, pain, overfull bladder and colon, fatigue, overtense muscles, illness and other disagreeable bodily states, etc. (Kretch, Crutchfield, and Livson 1969:498)

How strange that this introductory psychology text does not mention avoiding excess carbon monoxide. When a reserve is large enough, it remains hidden and we never think about it as a reserve at all. Indeed, our bodies lack effective regulatory mechanisms against carbon monoxide poisoning. We can't see it, smell it, or particularly feel it until it's too late. We have a *motive* for avoiding carbon monoxide, but it's never been much of a regulatory *problem* for us, so far.

When a reserve is so well hidden, a system seems able to function well for a long time without precise regulatory mechanisms, or apparently without regulatory mechanisms at all. Thus, the reserve may be either a "first resort" for rather crude systems or a "last resort" for systems with a sophisticated plait of regulatory mechanisms. Both the rich kid and the poor kid have reserves of cash, but the rich kid's first thought on *any* problem is to buy his way out of it. The poor kid, whose reserve is not nearly as great, has hundreds of tactics for solving personal problems without dipping into his "bail money."

By the Fundamental Regulator Paradox, the rich kid seems less likely to be *aware* of the regulatory power of money—until broke for the first time. But once the reserve is gone—and the shock is all the greater the less we suspected its existence—there is little that can be done. And yet, when there is nothing to be done, another strategy emerges from the *Lazy Law*—do nothing!

The "wisdom of the body" is such that the body is able, unlike some people, to "admit" that it probably had a mistaken image of the situation. After it has exhausted its repertoire of mechanisms and matters still seem to be proceeding from bad to worse, the body may simply "let go." In Selye's words:

. . . when disease has already become generalized anyway, it is in the best interest of the body not to put up defensive inflammatory barricades. (1956:120)

In other words, when all else fails, stop fighting and see what happens, lest you be destroyed by your own regulatory mechanisms.

For an individual, or a society, letting go is most likely to represent letting go of a worn-out model. The nervous breakdown, the revolutionary general strike, divorce, dropping out of school, these examples of letting go stir a ripple of fear within our breasts. We are like a man hanging by his very fingernails from the edge of a cliff, a cliff whose height he cannot see, but the image of which curdles his blood. He tortures himself until his fingers are torn to shreds, but finally he can take no more. Physically, he still has some strength, but he no longer has the will to stand the pain, so he lets go. And what happens? He finds that the ground is 3 inches below his feet!

The source of our fear is the general need to protect our images—our models of the world by which we regulate our lives. Because extraction of reliable models from the multivariable environment is such a long and expensive process, it normally pays a system to be conservative about preserving its model—even in the face of absolutely contradictory evidence. In science, the hard-won nature of the models is clear, and the necessity for preserving them in the face of evidence was codified into the *Law of Conservation of Laws* (Weinberg 1975:41). But this law is much more generally applicable than to the explicit models of science.

When Vickers spoke of "control by error in the second sense," he was speaking about controlling the error in our models. Just as the stability of linear systems could be pushed back one level to a question of stability of the matrix elements, so can the general question of stability be pushed back to the question of stability of the models behind regulatory behavior.

Verbal models—of science, religion, or magic—are rather exceptional among regulatory models in the ease with which they can be changed. Hume indicates this in his description of how numbers, the most codified of all our models, are most easily corrupted. Their very advantage as easily manipulable models gives them a sensitivity which the models of other systems cannot tolerate. That is why we write the amount twice on a check.

But even "errors" in language models are valuable, for "errors" can be a force for continuity as well as change.

Goody and Watt (1963) suggest that "the social aspects of remembering" contribute toward social integration and cultural persistence—most evidently in nonliterate societies which rely on oral tradition.

What the individual remembers tends to be what is of critical importance in his experience of the main social relationships. In each generation, therefore, the individual memory will mediate the cultural heritage in such a way that its new constituents will adjust to the old by the process of interpretation that Bartlett calls "rationalizing" or the "effort after meaning"; and whatever parts of it have ceased to be of contemporary relevance are likely to be eliminated by the process of forgetting. . . . The social function of memory—and of forgetting—can thus be seen as the final stage of what may be called the homeostatic organization of the cultural tradition in nonliterate society. The language is developed in intimate association with the experience of the community, and it is learned by the individual in face-to-face contact with other members. What continues to be social relevance is stored in the memory while the rest is usually forgotten: and language—primarily vocabulary—is the effective medium of this crucial process of social digestion and elimination which may be regarded as analogous to the homeostatic organization of the human body by means of which it attempts to maintain its present condition of life. (Goody and Watt 1963:307–308)

The Road Ahead

"Living backwards!" Alice repeated in great astonishment. "I never heard of such a thing."

"—but there's one great advantage in it, that one's memory works both ways."

"I'm sure *mine* only works one way," Alice remarked. "I can't remember things before they happen."

> Lewis Carroll
> *Alice in Wonderland*

We've now come to the end of our broad study of regulation. Now that we know generally why things remain the same, we are almost ready to study *adaptation*—why things change.

For one thing, we now understand that the study of adaptation may be conceived of as the study of *models* and how they change. But because of the chameleon nature of verbal models, we will find it necessary to separate the study of adaptation into two parts. On the one hand, we have "adaptics"—the adaptation of nonsymbolic systems. On the other, we have "symbolics"—the adaptation of systems with explicit structures for exploiting the flexibility of symbolic models.

In effect, the distinction between adaptics and symbolics is based on *continuity*—how strongly a model is tied to the past, or how easily it is changed. Because there is no sharp line between adaptics and symbolics, our next volume may more or less slide from one to the other. Or perhaps we'll decide it's better to pretend we *can* make a clear distinction. After all, we've often pretended in this volume that we could clearly distinguish between statics and adaptics.

At present, the successor(s) to this volume is merely a large file full of notes. Because we, like Alice, can't remember things before they happen, we can't promise just what the future work will be. Much will depend upon response to this work and its predecessor.

But before we set this work aside for the next, we must do one more job. Otherwise, we might leave too much noise in your models of regulation. We've tried to teach you how to think more clearly about regulation, but we've not prepared you for the hazards of unclear thinking, a constant danger in the general systems environment.

What we need is a reserve for you to fall back upon in case the models we've lent you don't seem to work. Naturally, more energy will be required to examine the noise in our own thinking processes, but with some help from Alice's Wonderland and Looking Glass friends, we hope to make the conclusion of this exercise amusing as well as educational.

QUESTIONS FOR FURTHER RESEARCH

1. Dog Training

One of the great obstacles in training dogs is the discrepancy, and potential clash, between the model of the dog and that of the human trainer. Language is so important in the human model that we tend to assume that it is the *only* signal in our communication with the dog. In fact, the word "Heel" may be simply noise to the dog, while the trainer's unconscious anticipatory leaning forward, just before pronouncing the word, may be the signal to which she responds. Observe yourself and others interacting with dogs. List as many nonlinguistic signals as you can, and test their signaling value by presenting them individually and watching for responses on the part of the dog. Now do the same for the trainer. Identify signals sent by the dog and responses made by the trainer. Does it matter that the dog is not "conscious" of sending signals?

2. Contests

In the United States, there are hundreds of sponsored contests each year based on "random selection" of winners. Consider the various ways in which such random selection contests might not be random at all, and how entrants could bias their chances favorably or unfavorably. Then compare your analysis with that in *How to Win Sweepstakes Prize Contests* (Danch 1971).

3. Social Psychology and Political Science

Machiavelli, in his classic primer for would-be leaders, set down three basic aspects of political functioning:

1. *Virtu*—the skill (virtuosity) of the politician.
2. *Necessitá*—the limits placed by the structure of the system.
3. *Fortuna*—events over which the human has no control.

Discuss this model of the political process in terms of the regulatory principles presented in this chapter. Discuss particularly the relationship between the personality of the politician and the noise he encounters in the course of his activities, and how various personalities differ in the way they distinguish *necessitá* from *fortuna*. (References: Machiavelli 1976; Christie and Geis 1970)

4. *Time Versus Information*

Scientific instruments may be designed for taking advantage of the repetition strategy in extracting weak signals from noise. For example, the tracking telescope permits the same region of the sky to be projected for hours on the same photographic plate by compensating for the turning of the earth, and the scanning electron microscope repeatedly scans an object with an electron beam. Discuss the design criteria for such instruments, and in particular how passive scanners—like the telescope—differ from active ones—like the scanning electron microscope—which can change the intensity of the probing beam. (Reference: Oatley 1972)

5. *On War*

Discuss the following remark:

In his magnum opus, *On War*, Clausewitz begins his analysis by an attempt to reveal the "true nature of war." He states in effect what war *would* be if it were not influenced by "extraneous factors," such as the time it takes to execute intended acts, imperfect knowledge of the state of affairs, human frailty, indecision, and so on. He lumped all these factors under the term "friction," an obvious allusion to the factors that concealed from superficial observation the underlying fundamental laws of motion. (Rapoport 1971)

6. *The Sporting Life*

The use of the lightest possible fishing line is the sporting way to catch fish because of artificial limitations placed on the methods or apparatus used. Choose your favorite sport—hunting, tennis, wilderness camping, baseball, or what have you—and make as complete a list as you can of artificial limitations placed on the participants to make the sport more "sporting" by leaving noise in the regulation process.

7. *Psychology*

Transactional analysts agree with general systems thinkers that the past is the best predictor of the future. They go even further and suggest that the future can be changed by changing the model of reality:

Script analysis can be called a decision theory rather than a disease theory of emotional disturbance. Script theory is based on the belief that people make conscious life plans in childhood or early adolescence which influence and make predictable the rest of their lives. Persons whose lives are based on such decisions are said to have scripts. Like diseases, scripts have an onset, a course, and an outcome. Because of this similarity, life scripts are easily mistaken for diseases. However, because scripts are based on consciously willed decisions rather than on morbid tissue changes, they can be revoked or undecided by similarly willed decisions. (Steiner 1974:28–29)

Consider further the implications of applying a decision model, rather than a disease model, to emotional disturbance. Think about which of the two models is more susceptible to failure—whether through its own success, or noise, or the ease with which it might be discarded if necessary.

8. *Introspective Psychology*

Relate an example from your own life where, after tortured resistance, you "let go" of some beloved model of the world. Describe the increase in discomfort and how it was rationalized, the event which provided the last straw, the immediate feelings on letting go, and the long-term effects on your life.

9. *Insurance*

The entire insurance industry is based on noisy models. Insurance represents a number of systems joining together to create a buffer that none of them could afford individually, knowing that *statistically* it will suffice.

One consequence of insurance is predicted by the Fundamental Regulator Paradox: when regulation through insurance is effective, then there will be little or no improvement in regulation of the risks themselves, unless, of course, the insurance company uses a rating system.

Discuss the impact of malpractice insurance on medical practice. Discuss the impact of fire insurance on fire. Discuss the impact of life insurance on life expectancy. Discuss what makes these three cases different.

15

Making Regulation Mysterious

The Hatter opened his eyes very wide on hearing this; but all he *said* was, "Why is a raven like a writing desk?"

"Come, we shall have some fun now!" thought Alice. "I'm glad they've begun asking riddles—I believe I can guess that," she said aloud.

"Do you mean that you think you can find out the answer to it?" said the March Hare.

"Exactly so," said Alice.

"Then you should say what you mean," the March Hare went on.

"I do," Alice hastily replied; "at least—at least, I mean what I say—that's the same thing, you know."

"Not the same thing, a bit!" said the Hatter. "Why, you might just as well say that 'I see what I eat' is the same thing as 'I eat what I see'!"

"You might just as well say," added the March Hare, "that 'I like what I get' is the same thing as 'I get what I like'!"

"You might just as well say," added the Dormouse, which seemed to be talking in its sleep, "that 'I breathe when I sleep' is the same thing as 'I sleep when I breathe'."

"It *is* the same thing with you," said the Hatter, and here the conversation was dropped. . . .

Lewis Carroll
Alice in Wonderland

The Hatter never explained why a raven was like a writing desk. The question remained a mystery to Alice and to millions of readers ever since. All the verbal games the Hare and the Hatter and the Dormouse played only served to distract Alice from the original riddle.

It's easy to get drawn into verbal games. They're fun. They're philosophy. But practical people, like Alice, soon lose patience with what seemed pure fun at the outset. Systems thinking has, since its inception, been a playpen of verbal games. Strewn about the floor of the pen are all the favorite toys of philosophy, such as the nature of mind, the purposefulness of the universe, and the possibility of perfection.

In this chapter, we shall begin our exploration of fallacious and unproductive thought about regulation by examining some of these philosophical toys. We'll play with them for just a bit, but we'll drop them, like the riddle, without any final answers. If you like riddles, you may like this chapter, but if you think we can *ever* find out the answer to them, perhaps you'd better skip ahead.

The Impression of Intelligence

"If I'd been the whiting," said Alice, whose thoughts were still running on the song, "I'd have said to the porpoise, 'Keep back, please! We don't want *you* with us!'"

"They were obliged to have him with them," the Mock Turtle said, "No wise fish would go anywhere without a porpoise."

"Wouldn't it really?" said Alice, in a tone of great surprise.

"Of course not," said the Mock Turtle. "Why, if a fish came to *me* and told me he was going on a journey, I should say, 'With what porpoise?'"

"Don't you mean 'purpose'?" said Alice.

> Lewis Carroll
> *Alice in Wonderland*

Professor Brunetti of New York University was one of the great teachers we had the privilege to know—a man who could teach any language to anyone, even a systems thinker. On the first day of each semester, he began by explaining what he would do if by some chance he ever became Dictator of the United States. He would pass one law, he said, and then immediately abdicate. That law would banish pencils, pens, and all other forms of note-taking instruments from classrooms, so that, finally, students would pay attention to what he was saying.

Those of us who teach systems thinking get a similar feeling about pencils, but there's something even worse, something that remains an impediment to generation after generation of would-be systems thinkers. Were *we* to be empowered to pass one law, it would banish the words "purpose" and "intelligence" from use in all systems writing and discussion.

Brunetti's Edict would interrupt the flow by which words are transmitted from the professor's notes to the student's notes—without passing through either brain. Weinbergs' Edict would interrupt a similar flow of empty words—and take about half of all systems writing with it. Gone would be the arguments "proving" that system X is not "really" regulating because it is not "intelligent," or has no "purpose." Gone, too, would be the counterarguments "proving" why system X actually *is* "intelligent" and *does* have a "purpose."

In case you've been asleep for the past 20 years, let's illustrate what we mean with a few examples:

Let us look at the "block diagram" of a large-scale computer. We have a system of rectangles, each of which represents some part of the machine. Some of these rectangles are labeled "program," "addition," "multiplication," "memory," and so on, and a system of arrows indicates the flow of information from one part of the machine to another part. There is no block labeled "imagination" or "unprogrammed comparisons." These essential features of creative thinking are totally missing in the machine. The machine follows a given program and does nothing else. (Brillouin 1968:165)

. . . A whole of which the parts are alive cannot, in its general characters, be like lifeless wholes. (Spencer 1904)

In view of the above considerations, I now suggest the following as necessary and sufficient conditions to be fulfilled, or to be assumed to be fulfilled, in order appropriately to regard any given behavior pattern as *purposive:*

There must be, on the part of the behaving entity, i.e., the agent: (a) a desire, whether actually felt or not, for some object, event, or state of affairs as yet future; (b) the belief, whether tacit or explicit, that a given behavioral sequence will be efficacious as a means to the realization of that object, event, or state of affairs; and (c) the behavior pattern in question. Less precisely, this means to say that the entity exhibiting that behavior desires some goal, and is behaving in a manner it believes appropriate to the attainment of it. (Taylor 1968:240)

The easiest way to answer these loquacious gentlemen is to draw the whimsical Figure 15.1. By using the Principle of Indifference, we can put any names on the boxes we wish, though names seem not at all a matter of indifference to these authors. Having satisfied their conditions with this WISE system, we turn their question back: With what porpoise?

Why have we occupied our valuable time drawing, labeling, and reading this inane diagram? When considering a nonhuman system, these philosophers want to peek inside the black box and see if it contains what they imagine their own heads contain. Brillouin imagines he has "imagination." Spencer feels the vibrations of "life." Taylor believes he holds "beliefs." And all three of them *know* themselves to be "intelligent."

Each requires that any *other* system, to win the honorific appellation

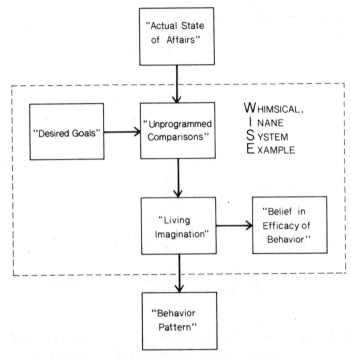

Figure 15.1.   By drawing these boxes and lines, and giving them the required labels, we satisfy the conditions of those who doubt the possibility of a non-human having "intelligence." Now, of course, they'll set other conditions.

"intelligent," must appear to him *from the outside* as he appears to himself *from the inside.* In other words, anyone else's black box must *do* whatever he merely *says* about his own white box. One wonders what each would think of the other!

It seems that only philosophers are granted the authority to prove things by verbal arguments alone. Computer programs, or other non-philosopher systems, must use both words and deeds. It's sufficient for a philosopher merely to *look* intelligent, but a computer program also has to *say* it's intelligent. But as any computer programmer knows, it's almost as easy to program a computer to say it's intelligent as it is to program philosophers to make the same claim.

It's hard to decide whether these utterances are senile maundering or immature egocentricity, but it's easy to see that they don't lead anywhere at all. But the rest of us shouldn't feel superior, for none of us seems sufficiently mature to miss a chance to take part in these debates. We feel impelled to waste our intellectual substance on refuting mysticism, which is nothing more than a ghost from our religious past.

Everybody is an expert in personal faith. We, too, can't resist the feeling of personal intelligence, and we imitate the mistakes of those we would refute.

The Myth of Superiority

"I declare it's marked out just like a large chess-board!" Alice said at last. "There ought to be some men moving about somewhere—and so there are!" she added in a tone of delight, and her heart began to beat quick with excitement as she went on. "It's a great huge game of chess that's being played—all over the world—if this *is* the world at all, you know. Oh, what fun it is! How I *wish* I was one of them! I wouldn't mind being a Pawn, if only I might join—though of course I should *like* to be a Queen, best."

Lewis Carroll
Through the Looking Glass

The fact of the matter is that we know no more about our intelligence or our purposes in life than a virgin knows about sex. We feel a few stirrings in our head—or in our stomach, if we grew up in certain cultures—and we read a few stirring words—"life," "purpose," "intelligence," "belief," "imagination," or "unprogrammed."

We are, thus, drawn into a peculiar form of the intellectual elitism which reaches its highest development among university professors. Though they may debate each other to the last split hair, underneath they agree about the beauty and profundity of the debate itself, and, by extension, the debaters. They look out upon their audience of students and stage whisper to each other, "Only you and I are intelligent."

It is this flattering elitism which draws us into stupid and wasteful arguments about purpose and intelligence. Because it appeals to a natural love of self, it easily insinuates itself into our thinking about diverse forms of regulatory mechanisms. We imagine that the ant is "smarter" than the grasshopper, for it remembers that hard-times follow easy and so stockpiles food for the coming winter—just as we salted away knowledge while those fraternity boys were having a good time. The society that lives by agriculture is "higher" than the one that lives by hunting. The one cultivates the soil in anticipation of a crop—just as we cultivate favor in the bureaucracy in anticipation of a promotion. The most "intelligent" person is the one who, like us, need not read the newspapers to know what will happen. She knows because of her "creative imagination."

In this manner, we are led by tiny steps to classify systems according to "intelligence." The plants are headed by the animals; the animals, by the primates; the primates, by people; people, by educated people; educated people by educated men; and educated men by *me,* if I happen to be a man. The progression must be modified a bit if I'm a woman, or uneducated, but the concept is the same.

This kind of ranking infests the social sciences like maggots in an organically grown peach. The social animals are headed by human societies; the hunting and gathering bands are headed by the agropastoralists; the agropastoralists are headed by the modern industrial state; and plunked right on the summit is *my* modern industrial state. No matter how fancy the terminological embellishment, the same phony argument peeps out, for example, in

One organic system is truly superior to another organic system if
(a) the first possesses every essential property of the second, but in addition to this relationship of essential similarity or kinship . . . with the latter, it possesses properties which are essentially lacking in the second, whereupon
(b) the total of these original properties of the first, with the preeminence of one of them, acts as the leading factor in its internal and external interconnections, that is, relates to the properties of the second system as major to secondary . . . (Kremyanskiy 1968:80)

Whatever the terminology, whatever the language, it's just another attempt to place a *value hierarchy* on systems. This may be good politics, or good philosophy, but it's irrelevant to good systems thinking. And besides, in these value hierarchies, guess who always emerges as *Numero Uno?*

Our best chance to escape this unproductive game is to avoid such emotionally charged phrases as "truly superior." As a substitute, we may tentatively offer the phrase "more complex"—with the understanding that complexity is certainly relative to an observer's capacity. Now we might make some progress in understanding regulators, by such an argument as the following. Every step by which a regulatory mechanism is removed from sensing or affecting the regulated variables requires the interposition of an additional model. Although the combination of two models *could* be less complex than either one, there is a tendency for complexity in models to rise as the time between sensing and acting grows. Seen another way, as the distance between cup and lip widens, there is greater opportunity for a slip, since continuity becomes less dependable.

Unless we are willing to invoke explicit magic, we must admit that the complexity built into the system's models has to be based on feedback from the past. If the accumulation of complexity in the model was a *long time* in the past, we may forget that accumulation and begin to imagine that the system is operating only on the basis of immediately available information. Such a system then *appears* to have some mystical ingredient not possessed by other systems in the same situation—intelligence, purpose, creative imagination, or just plain smarts. But since there is no mystery, making one only confounds our explorations.

Regulators do not require any mystical ingredient to work, even at the most complex levels. There *may be* systems with mystical ingredients, but science cannot study them. If a system gives us the *impression* of intelligence, we need not invoke an invisible hand which controls the market or an unseen eye which is on the sparrow. We may use the *metaphors* of intelligence—the "wisdom of the body"—but we are treading on the thin ice that separates the fresh air of science from the murky waters of metaphysics. As stimulants, these metaphors are indeed irresistible, but what they stimulate will usually be fruitless name-dropping. For heuristic purposes, it would be more intelligent to abolish "purpose" and "intelligence."

Searching for Nonexistent Set Points

"Would you tell me, please, which way I ought to go from here?"
"That depends a good deal on where you want to get to," said the cat.
"I don't much care where—" said Alice.
"Then it doesn't matter which way you go," said the Cat.
"—so long as I get *somewhere*," Alice added as an explanation.
"Oh, you're sure to do that," said the Cat, "if you only walk long enough."

Lewis Carroll
Alice in Wonderland

In cybernetics, the argument about purpose and intelligence has centered around the "set point"—naturally enough, in view of the evolution of cybernetics as the art and science of steering things. But any system which "walks long enough" is sure to get *somewhere*, so we cannot be certain that there *is* a set point. Because the set-point concept is useful, we are lured into using it as a heuristic in cases where no set point is ultimately found.

For example, one of the best regulated physiological variables is body weight. Although some of us have more trouble than others, most males between 25 and 55 vary in weight by less than 5 percent, except during an occasional illness from which they recover their weight in a few weeks. Considering the length of time—30 years—and the variety of environment—Mom's apple pie, Friday night fish fries, and football and beer weekends—one would certainly suspect that there is some set point against which body weight is measured. But there is no evidence that such a set point, in any explicit form, exists, nor does it seem likely that the body has a way of measuring its weight directly as it does brain temperature in the hypothalamus.

Nor is there any reason to believe there *should* be a set point—if we hadn't been immersed in cybernetic literature. Nor are the cyberneticists to blame if we cannot extend their models into regions they never attempted to explore. The search for a set point is a fine heuristic to add to Cannon's four propositions, but it lacks the strength of their generality. The conditions for stability simply do not require that there be an explicit place in the system on which one can put a finger and say, "Here is the point from which this system is controlled." Though many regulated systems will have such points, many will not, and there will not even be a sharp dividing line between them to help us know which case is at hand.

Take the case of the living cell, in which there seem to be thousands of regulatory mechanisms, mostly still unknown. An extra heuristic device would certainly be welcomed here, but the set-point concept is not it. In one simple type of feedback loop, illustrated for lactose in Figure 15.2, the presence of a "food" substance, lactose, somehow *induces* the lactose-breaking enzyme, beta-galactosidase to be produced (see Jacob and Monod 1961). Nowhere in the process does there seem to be any need

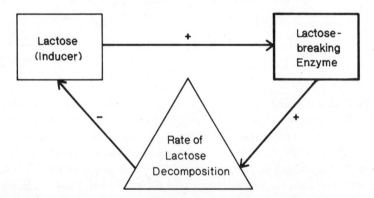

Figure 15.2. An enzyme is called "inducible" if the presence of the molecule it is to act upon *induces* the creation of the enzyme. This induction creates a stable feedback loop, but with no distinct set-point.

for a set point to make the process more effective. Fifty trillion bacteria can't be wrong.

Or consider another type of cellular regulation, the so-called "repression," as shown in Figure 15.3. Umbarger (1956) first showed this type of regulation in the synthesis of isoleucine (X in our diagram). Here, isoleucine is called the "repressor" because it *inhibits* an enzyme (threonine dehydrase) needed for the synthesis of the first substance (A) in a chain (A-B-C- . . .) leading ultimately to isoleucine. Here again, and in hundreds of other repression regulatory syntheses, stability is obtained without any hint of where a set point might be.

Rather than choose even more examples from molecular biochemistry, let's elaborate the point using a more familiar model, our imaginary population of Chapter 11. That population was characterized by a matrix, M, whose values were indicative of the influence each age component had on the size of each other age component some years later. In Chapter 11, with the matrix representing linear relationships among the components, we saw that stability of the population was a rather unlikely event.

Moreover, if stability did happen to occur, it was the result of the matrix itself being extremely stable, though we didn't know then why the matrix would be stable or unstable. Let's think now of a modification that

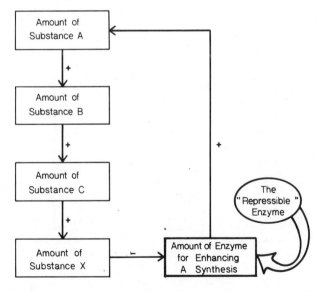

Figure 15.3. An enzyme promoting the synthesis of precursor (A) of substance X is inhibited by the presence of X, so little or no precursor is synthesized, which ultimately leads to a stable level of substance X.

might produce a more stable matrix, and thus, perhaps, a more stable population.

Suppose that in this society, new housing is scarce. Indeed, it is so difficult for young couples to get housing of their own that they often delay marriage and/or children for many years, until they can live separately from their own parents. Imagine, too, that very little new housing is being built, and that most of the available space results from the death of old people.

In such a population, a drop in the number of old people will mean more space for new families, and therefore an increase in the number of children being born. In the matrix, the birth rate expressed as a fraction of the young adult population is the element M_{12}. Schematically, this effect of the number of old people on the birthrate can be represented as in Figure 15.4.

The system shown in Figure 15.4 is no longer a simple linear system. The multiplication sign at the junction of the two lines heading for the 0–14 box shows how the number of births is computed—by multiplying the birth *rate* (M_{12}) by the *number* of young adults (V_2). This is the same computation we made in the linear system, but in that case, M_{12} did not change, so the number of births was a constant (M_{12}) times a variable, V_2. Now, however, M_{12} changes according to the value of V_3, so the number of births is determined by the multiplication of *two* variables—a nonlinearity.

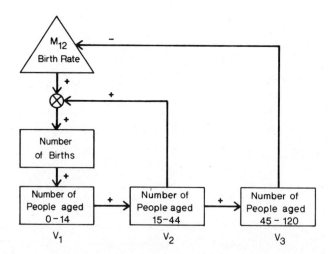

Figure 15.4. Control of birth-rate without set-point. A more stable population results when an excess of old people has a depressing effect on the birth rate, M_{12}, resulting in a non-linear system with no explicit set point for total population.

We can determine the *qualitative* effect of this change in the usual way, by marking each effect plus or minus and projecting the signs onto the diagram of possible effects. The inner loop is a typical positive feedback loop of a potential "population explosion"—runaway feedback. The outer loop, however, is different. Forty-five years after the 0–14 age group increases, the young whippersnappers have become old fogeys. The increased number of old people then has a depressing effect on the birth of more 0 year olds. In other words, this self-regulating population ultimately creates its own stability through the relationship between the number of old people and the production of new members of the society.

Such a population should level off somewhere, unlike the earlier model which kept running away unless precise control was kept on the matrix, M. If we look at Figure 15.5, we see that our expectation is fulfilled. The system is started in the state

$$(900, 300, 200)$$

goes through some rather complex oscillations, and eventually settles down, more or less, around

$$(100, 150, 250)$$

The stability of the model can be tested by trying a variety of other starting populations, each starting with the same matrix, M. In Figure 15.6, we see how the system behaves when started at

$$(200, 400, 900)$$

Again there is a series of oscillations, but quite different from the ones of Figure 15.5. When the system finally settles down, the population is at about

$$(500, 800, 1000)$$

Thus in the first case the population started at 1400 and eventually stabilized at about 600; while in the second, the population started at 1500 and stabilized at about 2300.

Now, if there were a set point *inside* this system, that would mean it was either in M or on V_0 (the initial population). It clearly can't be in M, because M was the same in both Figure 15.5 and Figure 15.6, yet the stable population was entirely different. Therefore, the set point must be in V_0, but where? Well, yes, it's *there*, but not in any place you can put your finger on, or in some dial a government official can twist to get some desired population. In fact, nobody can influence it at all, in the kind of

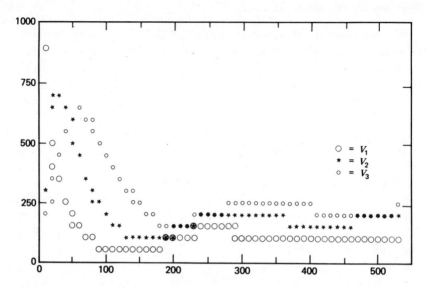

Figure 15.5. The behavior of the population shown in Figure 15.4 starting with a population of (900, 300, 200) and oscillating until it stabilizes around (100, 150, 250).

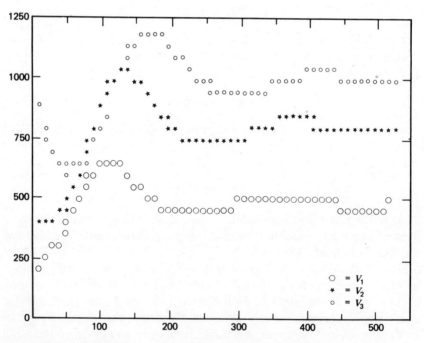

Figure 15.6. The behavior of the population shown in Figure 15.4 starting with a population of (200, 400, 900) and oscillating around (500, 800, 1000).

society we know, because we can't just kill people or move them around willy-nilly to achieve some desired level of future population.

When dealing with an actual system, as a *member* of the population under study, the whole concept of a set point takes on some strange attributes. We don't have the privilege of running our system as many times as we like on the computer, to see what causes the equilibrium to come out differently. In fact, we see but one outcome, the outcome we call "history." Because it's the only outcome we have, and because we are part of it, we are particularly fond of it, and can't believe it came out that way just because it had to get *somewhere*.

If we are to carry on in the face of life's insults, we have to believe, somehow, that we are controlling our fate. And so we begin searching for a set point—armed with a generous government grant. Perhaps the only more futile search is for a set point in the stock market, a game which millions play every day. Surely the Stock Market cannot just come out *somewhere?* Surely the level on Wall Street is set by the belief that people have about the belief that people will have about the level on Wall Street? So if everyone *believes* there's a set point, then there will be one, so let's leave it at that.

Other Set Point Fallacies

Alice watched the White King as he slowly struggled up from bar to bar, till at last she said, "Why, you'll be hours and hours getting to the table, at that rate. I'd far better help you, hadn't I?" but the King took no notice of the question: it was quite clear that he could neither see her nor hear her.

So Alice picked him up very gently, and lifted him across more slowly than she had lifted the Queen, that she mightn't take his breath away; but, before she might put him on the table, she thought she might as well dust him a little, he was so covered with ashes.

She said afterwards that she had never seen in all her life such a face as the King made, when he found himself held in the air by an invisible hand, . . .

Lewis Carroll
Through the Looking Glass

Belief in a chimerical set point is but one facet of the set-point fallacy. Another facet is the belief that a real set point doesn't exist, when it does.

In human affairs, one group may not want another to know of the existence of a set point. We read a pious announcement that the government will set no quotas on hiring the various races, creeds, colors, sexes, or ages. After all, bleat the politicians, a quota, or set point, wouldn't be *fair*.

But we need not believe the politicians, if we don't care to. When we find that an entire agency has not a single woman, or a single Native American, or never hired anybody over 45, we can assert that there *must* be a quota. The set point is set at zero, from a political action point of view. We 50-year-old Native American women are certainly not going to give up trying to pressure them into hiring us merely because some bureaucrat says there's nothing in writing. We believe there's a set point, so there *is* a set point, and we take action accordingly.

The Wall Street speculator and the activist can afford to act as if there is a set point, or isn't a set point, merely on the basis of their beliefs about the matter. For the scientist, the situation is different. Her experimentation will take one course if she believes in an explicit set point; another, if she doesn't. Consider the following statement:

It is therefore tempting to conclude that there exists a regulatory system controlling body weight, or some correlate of body weight. The regulated variable has to be related to body weight and probably not to food intake. This concept of "ponderostat" implies that the regulated variable, whatever it may be, is compared to its set value. The difference between the two will judge the internal input after food intake and in turn, . . . will control food intake. (Cabanac 1971)

Cabanac, in other words, is pursuing the possibility we rejected earlier—that body weight is regulated by a set point. Given the one view, he is sure to press the investigation. Given the other, we are sure to seek other intellectual pastures in which to graze. Who will prove right is not a matter to be settled by general systems thinking, but by experiment. And it won't be *our* experiment, but Cabanac's, which seems only fair.

If we establish, beyond doubt, the existence of a set point, we are still not rid of fallacies. What, we must inquire, keeps the set point *itself* set and steady?

If there should actually be a "ponderostat," my diet shall prove, as it were, fruitless. As soon as I return to unsupervised eating, my system will pull me back up to my previous level as surely as it pulls back my heartbeat after a strenuous volleyball struggle in the pool.

If I wish to reduce and *stay* reduced, I don't want to know merely that there *is* a ponderostat, but also what in the world sets it? Even further, is my ponderostat set once, perhaps at birth, never to be changed by willpower or Weight Watchers? Or can it be set and reset, like a thermostat, to fit the fashionable look of today?

These questions are not nearly as silly as they sound. Frequently enough, some nonset level of stability is achieved and held sufficiently long for it to *become*, itself, a set point for other variables. If Eldon manages to get down from 250 to 175 and hold it there for six months, he may find his wardrobe investment is sufficient incentive to hold it there for a few more. Many less obvious changes may take place in his life style, and even his physiology, which may account for the "plateau" phenomenon dieters so often experience.

To take a more ancient example, the salinity of blood may have originally been related to the salinity of ocean water. Over millions of years, though, an organism must come to depend on a constant blood salinity in a thousand and one ways. When the salinity of the ocean changes—or when the organism leaves the briny deep for a landlubber's life—it will have to provide an explicit mechanism for setting and holding this constant level in its blood.

To cite another case, the price of milk may establish itself through an "invisible hand" of the market. After a considerable duration of steadiness, however, dairy farmers will have adapted in dozens of ways to that price level. Thus when the "invisible hand" draws back into its sleeve, there will be lobbying to replace it with a *seen* hand—a price support or subsidy. Though the initial price *may* have been arbitrary, without any set point whatsoever, the cost of arbitrary *changes* in this level swells to huge proportions once dependent mechanisms have been established.

Although economists may believe that the "invisible hand" is a set point, or point setter, practical business people have never had much faith—as opposed to ritual political worship—in *free* enterprise. The "invisible hand" sets the point where it wants, not where the entrepreneur wants. Consequently, practical people will always attempt to consolidate gains by establishing a set point that *can* be seen, and set, by them. Any free market is thus the most ephemeral of phenomena, soon to be replaced by cartels, monopoly, trade associations, fair trade agreements, tariffs, quotas, cliques, inner circles, trusts, syndicates, pools, consortiums, regulatory agencies, exclusive territories, and even trade unions. The business of business is far too important to be left in "invisible hands."

The Quest for Perfection

"You see," he went on after a pause, "it's as well to be provided for *everything*. That's the reason the horse has all those anklets round his feet."

"But what are they for?" Alice asked in a tone of great curiosity.

"To guard against the bites of sharks," the Knight replied. "It's an invention of my own."

Lewis Carroll
Through the Looking Glass

Try as they might, the capitalists, the socialists, or even the communists, never quite succeed in making the market behave the way a proper market *should* behave. Perfection is their common goal. Disappointment is their common fate.

Knights frequently symbolize the quest for perfection—the Holy Grail, the chaste princess, even the perfect society pursued by the Knights of Labor. Seldom are the consequences of that quest portrayed so succinctly as in the image of the White Knight. Laden with devices for warding off shark bites and other imagined threats, he is unable to guide his horse through the execution of the simplest maneuver. Is there a lesson here for those who would guide the unseen hand? To paraphrase the Lump Law, in trying to regulate everything, we regulate nothing.

Though for a variety of reasons, including the Regulator Paradox, perfect regulation is not possible, we persist in the quest of this Holy Grail. Some commentators have suggested that the White Knight is Carroll himself, but there is a smidgin of him in each of us. If perfect regulation were like the perfect cup of coffee, scant harm could come from this vain quest for flawlessness. In complex systems, it courts catastrophe.

By expending huge amounts of energy, we may bring a variable under perfect control—for a limited time. As time passes, however, the knowledge of the system's and the environment's characteristics—the model—becomes less and less reliable. Thus when a really critical moment arrives, the system is apt to behave in some singularly inappropriate way.

Some years ago a grand campaign was launched in the United States to control forest fires, to "stamp them out" completely. In the past, attention had been devoted merely to limiting the damage from those fires that occurred—less than perfect regulation. The more stringent quest was so successful that in many forests, long periods elapsed with no fires at all. But forest fires not only destroy forests; they test the resources and methods of fighting other fires. After a forest has not burned for many years, the accumulation of combustible material gets so great that a fire in it will have different characteristics than it might have had earlier, and the old fire-fighting techniques may no longer be adequate—and nobody knows about it. Result: the fire destroys not just the underbrush, deadwood, and some smaller trees, but the very forest itself.

Such disasters are common enough. We get them artificially when we raise chickens in germ-free environments, and naturally when we raise children in environments free of candy, alcohol, or sex. What they have in common is the failure to recognize that *purity is the enemy of learning*. With children, all but the most extreme cases eventually recover from a puritanical household. The others become great writers or are cast into asylums, or both. But the larger the scope of our puritanical efforts, the grander the scale of the eventual denouement.

A tragic example is the natural history of poliomyelitis. Before a society reaches the great sanitation revolution, polio will be endemic and rare, for the persistent contamination of the environment with human wastes assures that only a few children escape exposure in infancy. In infancy, the disease is generally not serious, and has the benefit of conferring life-long immunity from further attacks, an immunity which may even be transferred to the unborn. As a society begins to cleanse itself, however, more and more children become young adults without having the good fortune to contract polio. Should they then be unlucky enough to be ex-posed as adults, to a crippling strain, consequences are cruelly severe. Moreover, since there are few immune persons, cases may spread in waves, or epidemics. Such is the possible outcome of the quest for per-fection.

The costs of perfectionism are not necessarily so pathetic, but there are always costs. Even when we are not trying for perfect regulation, there are costs for just trying to do a little better. The obverse of this relation-ship is the savings that can accrue to a relaxation of an attempt to regu-late some variable too strictly. For instance, in international finance, it is generally considered desirable to have stable exchange rates among cur-rencies, if only to discourage ruinous speculation. Therefore, central banks would intervene when, say, the United States dollar deviates from the Swiss franc by more than the specified range, such as 1.74–1.78. By "widening the gold points," as proposed by Keynes in the 1920s, the cen-tral banks should be able to achieve the desired stability with less fre-quent and less massive intervention. In fact, since central banks may sim-ply be too ponderous to act as swiftly as speculators, such a widening of exchange ratios may be the only feasible way to keep the system from shaking itself to pieces.

Yet in spite of all these experiences, the quest for perfect regulation dies hard, particularly in political and religious literature. Or if it dies, it seems to be resurrected in new forms. We cannot, it seems, perfectly regu-late it out of existence.

As an example of the quest in another costume, consider the following quotation, in which Polanyi argues that the deviation from perfect regu-lation is not bad in itself, but as the potential nucleus of an inevitable positive feedback loop.

Even an organized balance-of-power system can ensure peace without the permanent threat of war only if it is able to act upon these internal factors directly and prevent imbalance *in statu nascendi.* Once the imbalance has gathered momentum only force can set it right. It is a commonplace that to insure peace one must eliminate the causes of war; but it is not generally realized that to do so the flow of life must be controlled at its source. (Polanyi 1957:9)

One of the experiences that keeps this fallacy alive is the observation that small disturbances often grow into civil wars, revolutions, or major heresies. Yet it is not generally observed that in many of these cases, the positive feedback loop involves the very efforts to control. When you are carrying a large stack of plates across a room and the stack begins to wobble, the best you can do is keep walking. Any energy you try to put into rectifying the stack just amplifies the process of derangement, so that eventually the plates are thrown all about the room.

In the modern theory of guerrilla warfare the insurgents live off the very armaments poured in by the established government to suppress them. In religious heresies, only the established church has a wide enough information network to broadcast the blasphemous words. It is a problem of scale—the same principle that gives the tiny judo expert the ability to break a huge opponent's arm. Without the weight and thrust of the opponent, the expert lacks the strength to cause real damage—it is the system itself that has the energy to destroy itself.

To be sure, great cataclysmic events often start with some tiny nucleus—a cow kicks over a lantern, someone shoots an archduke, or a piece of paper is nailed to a church door. But the nucleus has no meaning without the massed accumulation of stress waiting to be released. Tying up all cows' legs is not going to prevent another Chicago Fire: that will only be done by eliminating the conditions for combustion. If it hadn't been Mrs. O'Leary's cow, it would have been Mr. McGoff's cigar. In the whole world of human affairs, there is always another nucleus ready to step forward when the tinder is dry.

If the conflagration does not come, it is not for lack of nuclei, but because of the normal regulatory powers that govern our affairs. As Locke observed:

Such revolutions happen not upon every little mismanagement in public affairs. Great mistakes in the ruling part, many wrong and inconvenient laws, and all the slips of human frailty will be borne by the people without mutiny or murmur. (1947:235)

Locke knew that no society could function if it reacted violently to every wind of change, not to speak of breeze. Still, there do come times such as that so aptly described by one of the Kadet party leaders just before the Russian Revolution:

We are treading a volcano . . . The tension has reached its extreme limit . . . A carelessly dropped match will be enough to start a terrible conflagration . . . Whatever the government—whether good or bad—a strong government is needed now more than ever before.

The danger is real enough, but strong government is not the remedy. If conditions are right for an explosion, the attempt to nip each and every nucleus "*in statu nascendi*" is doomed to fail. Should the government lose its nerve—bolster the corps of secret police, provocateurs, and spies—its every move not only fails to eliminate nuclei but also increases the tension. When will governments learn that the presence of secret police among their own people is one of the surest ways of fanning small sparks of discussion into great flames of revolution? Only, perhaps, when they become perfect.

QUESTIONS FOR FURTHER RESEARCH

1. *Intelligent Machines*

Although we disapprove of spending time in counterargument to those who argue against intelligence in machines, we cannot resist calling attention to one of the shortest, neatest, and sweetest put-downs in the whole business. In two pages—which is about all the subject deserves—Paul Armer (1963) manages to examine and demolish by ridicule no less than seven forms of argument against machine intelligence. But these arguments can be turned around. Study Armer's article and see how the seven arguments could be used in trying to decide the question: "Can people think?" What conclusions can you draw from this exercise?

2. *Anthropology*

Anthropologists have not entirely escaped the traps in which their fellow social scientists have been caught. They too have classified the world into "culture types" existing along a continuum, based on relative complexity, ranging from the "simplest" band societies to the "most complex" modern states. Morton Fried (1975) has exposed the trap by writing about "the myth of tribe." He explains how the "tribe" was "manufactured" by powerful conquerors whose style of absentee administration required that the conquered peoples be treated as an undifferentiated unitary group. Far from being a genuine and indigenous organizational form of society, "tribes" are nothing more than "products and servants of the state." Read Fried's article, and think about various "tribal" societies you have studied. What would be the consequences for anthropology of rethinking the concept of "tribe" and, for that matter, of other hierarchical constructs in the discipline?

3. *Optimization*

Mathematics has qualities that appeal to people who are not immune to the quest for perfect regulation. When we combine mathematics with the search for perfection, we get optimization theory. If you have experienced the mathematical elation of studying optimization, try to relate your feelings to the pitfalls that have greeted other attempts at perfect regulation. In what ways does modern optimization theory avoid these pitfalls? In what ways does it repeat the selective blindness of its predecessors? (References: Cannon, Cullum, and Polak 1970; Sworder 1966)

4. *Economics*

List as many mechanisms as you can which tend to regulate the level of the United States economy. Classify them into explicitly recognized regulators of the economy, such as government spending on certain public works or deliberate tax cuts; explicitly recognized regulators of other economic variables that happen to regulate spending, such as unemployment compensation and farm aid; explicit laws that regulate the economy implicitly, such as laws concerning gambling or abortions; and nonjural regulators, such as preference for clear water or large families. How many mechanisms would you guess there are that are not on your list?

16

Overly Simple Views of Regulation

"Only it is so *very* lonely here!" Alice said in a melancholy voice; and, at the thought of her loneliness, two large tears came rolling down her cheeks.

"Oh, don't go on like that!" cried the poor Queen, wringing her hands in despair. "Consider what a great girl you are. Consider what a long way you've come to-day. Consider what o'clock it is. Consider anything, only don't cry!"

Alice could not stop laughing at this, even in the midst of her tears. "Can *you* keep from crying by considering things?" she asked.

"That's the way it's done," the Queen said with great decision: "nobody can do two things at once, you know."

Lewis Carroll
Through the Looking Glass

The way some people carry on, you would think the White Queen was right. But people can do several things at once, and so can regulators. Sometimes those things complement one another, and sometimes they even work opposite to the way they seem to be working. In this chapter, we'll play with some examples of other thinkers who seem to agree with the Queen.

Plait Fallacies

> "Come, tell me how you live," I cried,
> "And what is it you do!"
>
> He said, "I hunt for haddocks' eyes
> Among the heather bright,
> And work them into waistcoat buttons
> In the silent night.

And these I do not sell for gold
Or coin of silver shine,
But for a copper hapenny,
And that will purchase nine.

"I sometimes dig for buttered rolls,
Or set limed twigs for crabs;
I sometimes search the grassy knolls
For wheels of Hansom-cabs.
And that's the way" (he gave a wink)
"By which I get my wealth—
And very gladly will I drink
Your Honour's noble health."

Lewis Carroll
Through the Looking Glass

We know from Cannon's principles that eccentric old men aren't the only systems that "comprise a number of cooperating factors brought into action at the same time or successively." When the factors are successive, we may notice only one at a time, which could lead us to the erroneous belief that "What is it you do?" can be answered in the singular. If we understand *why* regulatory mechanisms may be successive, with one taking over when the other can no longer cope, we may be able to avoid the kind of simplistic thinking that has led to so many disasters in systems designed by people, rather than evolved by nature.

Let's carry our thoughts about the design of a flush toilet a few steps further. Suppose our tank toilet has been regulating its water level perfectly for years, but now begins to experience erosion of the rubber gasket on the valve which stops the inward flow of water. Such erosion— the gradual giving way of one of the underlying aggregate regulators—is ubiquitous in systems that display intermittent mechanical motion like the opening and closing of the valve. Indeed, that is why a *replaceable* gasket is used—to minimize repairs when one wears out. (Figure 16.1)

But let's carry the design thinking to its natural conclusion. If the gasket is designed under the assumption that it will eventually wear out, how is the float valve regulator going to function when that happens? Water will trickle into the tank even when the valve is "closed" by the float reaching the set point. This creates an input with no corresponding output, unless the toilet is flushed very frequently. Thus, in a busy environment, the system will actually be helped in its job of regulating, but when the bathroom traffic slows, water will rise above the set point, eventually

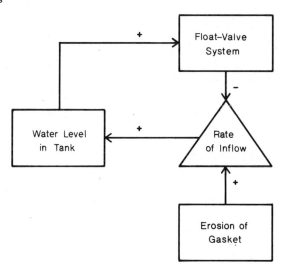

Figure 16.1.   A toilet tank with an eroded gasket on the input valve, leading to a condition in which the water level eventually goes over the top of the tank.

spilling over the top of the tank and onto the floor. We can imagine how puzzling a toilet such as this would be to its custodian. It might appear trouble-free during the week, only to overflow on the weekend when the building is empty.

With modern toilets, this disaster is prevented by the addition of a supplementary regulator, shown in Figure 16.2. An overflow pipe with an open top below the top of the tank allows the excess water to spill into the pipe before it gets high enough to spill over the top of the tank. Naturally, we put the top of the pipe above the set point of the float valve. Otherwise water would flow down the overflow even when the gasket is in perfect condition.

Figure 16.3 shows the block diagram of the toilet with this pair of regulatory mechanisms. Examining the diagram, a natural question is, "Why do we need to retain the float valve regulator, once the overflow pipe has been added?" Certainly the overflow is a simpler, almost foolproof, mechanism for preventing water from reaching the top of the tank. It will allow the tank to refill to a high level after a flush, without the complexities and uncertainties of the float valve. Why not get rid of the float valve?

From the point of view of *reliable* operation, there is little reason to retain the float valve once the pipe is installed. From the point of view of *efficient* operation, there may be a big reason. Once the tank has been refilled, the overflow pipe would keep the level constant by allowing all the incoming water to flow out the drain, thus wasting a great quantity of

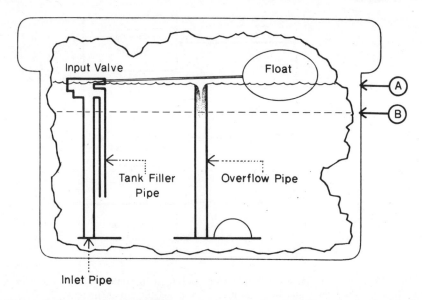

Figure 16.2. A toilet tank with two mechanisms for controlling the level of water in the tank, the float valve regulator and the overflow pipe.

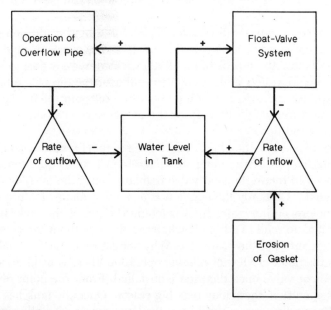

Figure 16.3. Diagram of effects for the toilet tank of Figure 16.2, with a leaky gasket on the input line.

294

water. If water is plentiful, this inefficiency may be of little consequence, but in a general-purpose design, it wouldn't be satisfactory.

The waste could be reduced by running the inflow very slowly, but then the toilet would have a very slow recovery rate. This design might be good in a remote place where there was a small, steady flow of water and not much toilet traffic. But to get a quick recovery system that is also economical of water, we use the float valve, backing it up with the over-flow pipe to guard against corrosion or other float valve malfunctions. Though we don't want overflow all the time, we're more than happy to waste water on occasion to save mopping the floor.

The toilet is a *designed* system, so the "purpose" of the two mechanisms is clearcut. Simplifying the environment to two variables—cost of water and frequency of use—we can then visualize the contribution of each mechanism to the overall problem of stability. In Figure 16.4, the shaded region shows that region of the environmental state space in which the overflow design would be acceptable. If water cost is very low, the over-flow method alone is all we need. As water cost rises, we can get by with the overflow alone if use is infrequent, so we can trickle the water in without worrying about fast recovery.

In Figure 16.5, the shaded region shows that region of the environmental state space in which the float valve design would be acceptable. When

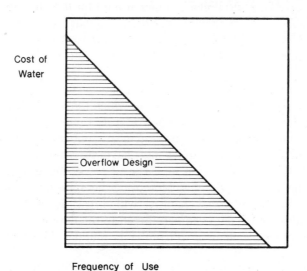

Frequency of Use

Figure 16.4. The region of a two-dimensional environmental state space in which a simple overflow toilet would give satisfactory economic performance (shaded), either because water is cheap or usage is low.

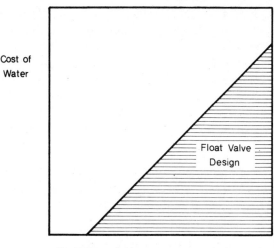

Cost of
Water

Float Valve
Design

Frequency of Use

Figure 16.5. The region of a two-dimensional environmental state space in which a simple float valve toilet would give satisfactory economic performance (shaded), either because water is cheap or because enough people use the system to make a long, undetected overflow unlikely.

frequency of use is very low, we run the risk of an expensive disaster if the float valve fails and nobody comes around for days to notice the overflow. At some point, usage is high enough so that we can risk occasional failures, if water is cheap enough. When there is much traffic, we can be sure that *somebody* will notice before there's too much water over the dam.

In Figure 16.6, we see how combining the mechanisms enlarges the region of the environment's state space in which the one design can operate economically. At very low frequency of use, the extra cost of the additional mechanism may make this system uneconomical, and at high frequencies of use, no single toilet may have adequate capacity. For the most part, though, the region includes all the economical operating regions of the other two designs. Moreover, there are many circumstances in which one of the two mechanisms will "break down" for a variety of unpredictable causes not even considered in the design—circumstances in which the other mechanism will probably muddle through until repairs are made.

We see, then, that even in "designed" systems, the "purpose" of each mechanism is not necessarily clear. In some environments, the overflow pipe backs up the float valve; in others, the float valve essentially backs up the overflow pipe. And in many cases, one backs up the other in ways the designer never anticipated. In "natural" systems, it becomes even less

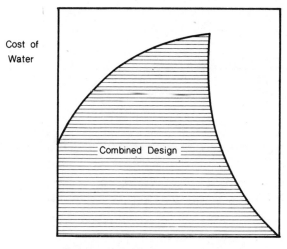

Figure 16.6. The region of a two-dimensional environmental state space in which a combined float valve and overflow pipe toilet would give satisfactory economic performance (shaded).

clear which is the "backup" mechanism for the other, as we can illustrate with a return to our little simulation of a village population.

In the system modeled in Figure 15.4, the active regulation was entirely through control of births, with deaths assumed to be a constant proportion of each age bracket. We know, though, that death can be nonlinear, too, as when some endemic disease becomes more frequent when conditions are crowded. Suppose this *density-dependent* disease factor influences infant mortality—the value of M_{11} in the structure matrix.

Figure 16.7 diagrams the effects in this modified village system. The heavy line surrounding the three population segments creates a block representing the population as a whole, because it is this population, rather than any segment, that determines the overall density, or crowding. The line from this block to survival rate, M_{11}, indicates that it is indeed the overall population density that affects infant mortality. By checking the loop from "number of people aged 0–14" to "overall population" to "survival rate" and back to "number of people aged 0–14," we see that this density-dependent infant mortality potentially contributes to stability of the overall population.

Figure 16.8 shows a computer simulation of this new system starting in the state

$$(200, 400, 900)$$

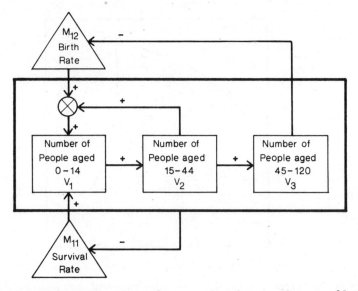

Figure 16.7. A density-dependent influence on the infant mortality rate adds another potentially stabilizing feedback loop to the village population model.

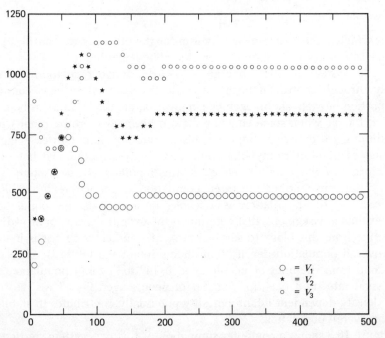

Figure 16.8. The behavior of the village population model of Figure 16.7 starting at (200, 400, 900) with an initial birth rate of 0.40. The final birth rate stabilizes at 0.34.

the same starting state as Figure 15.6. Comparing these two figures shows that the additional mechanism has somewhat smoothed out the extreme oscillations seen in Figure 15.6 and has brought the population to an almost constant level rather more quickly.

But comparing just two simulations hardly tells the whole story of the difference in stability between the one-mechanism and two-mechanism systems. The new system will be more stable even in the face of a variety of structural changes. It is not just more stable, it is more stable *structurally*.

To take just one example, if housing suddenly became very cheap, the regulatory ability of the birth-rate loop would be seriously reduced. Both systems would rise to new levels of population. The one with the addition of infant mortality would keep the rise slower and smoother, giving the system much more chance to develop other reactions to the population explosion.

The One-to-One Fallacies

"When *I'm* a Duchess," she said to herself (not in a very hopeful tone, though), "I won't have any pepper in my kitchen *at all*. Maybe it's always pepper that makes people hot-tempered," she went on, very much pleased at having found out a new kind of rule, "and vinegar that makes them sour—and camomile that makes them bitter—and—barley-sugar and such things that make children sweet-tempered. I only wish people knew *that:* then they wouldn't be so stingy about it, you know—"

Lewis Carroll
Alice in Wonderland

Alice had hit upon a new kind of thinking that worked so well for pepper that she applied it to all sorts of common substances. One ingredient, one effect—what could be simpler? Had Alice been talking about regulatory mechanisms and the variables they regulate, she would have been part of a crowd.

The general advantages of a *plait* of regulatory mechanisms encourage continued probing for more than a single source of well-regulated behavior. Even so, people persist in the fallacy that mechanisms and variables are in one-to-one correspondence. This simplistic idea is *so* wrong that it fails both ways:

1. Missing a second regulator for a single variable.
2. Missing a second variable affected by a single regulator.

Nowadays, we hear many arguments about "energy" and systems. The idea of "energy" as the universal regulator of all systems was wonderful in physics, as far as it went, but often contains the seed of one-to-one thinking. In social sciences, the same kind of universal elixir property is often attributed to "money," or something supposed to be closely related to money, like "power."

It's not always easy to resist these fallacious arguments, perhaps because of their intuitive appeal to our common experience at being overpowered by money. When the arguments are put forth by a prestigious figure and wrapped in obscure webs of syntax, we have almost no chance. Here's an example written by one of the big powers of sociology, held in universal awe because almost nobody understands him.

The money held by a social unit is, we may say, the unit's capacity, through market channels under given rules of procedure, to command goods and services in exchange, which for its own reasons it desires. Correspondingly, the power of a unit is its capacity, through invoking binding obligations (e.g., civic obligations such as military service, contractual obligations, the obligation to follow vested leadership, and so forth) to contribute to collective goals, to bring about collective goal-inputs that the "constituents" of the collective action in question desire. (Parsons 1964:46)

If general systems thinking does nothing else for you but help you slice baloney like this passage, it will be a resounding success. Put aside the grammatical error. Put aside the magnificent fog index (Gunning 1952) of 26, compared with, say, *Atlantic Monthly*'s 12 and *Time*'s 11. Consider only the fallacy of comparing "power" with "money." Parsons evidently thinks of "power" as a scalar (one-dimensional) quantity that can be summed from a variety of sources to obtain a single power "capacity." But "power" here is evidently a vector (multidimensioned) because "civic obligations" are varied and cannot be added together. For instance, changes in the social structure will have differential effects on the different components of this power vector. Consequently, a mechanism for regulating one component may not suffice for regulating another. Indeed, two components can easily conflict, as when entering military service prevents fulfillment of a contract.

Whatever idea Parsons was trying to communicate, it doesn't hang together when supported by the money analogy. The money analogy is particularly good at exposing latent one-to-one thinking because "money" itself can hardly be considered one-dimensional. We aren't speaking here of such curiosities as Mammy Yokum's favoring coins over paper, regardless of denomination. There are multitudes of real ways in which money doesn't act as a scalar quantity.

A dime is one-tenth of a dollar, but we may be more than willing to pay a dollar for a dime when we are under the influence of amoebic dysentery and faced with a pay toilet. On a larger scale, Gresham's Law deserved the name of Law only because it recognized that money was *not* scalar, in the face of the general belief that it was—thus "explaining" why with two kinds of coin in circulation, one "drives out" the other. And, contrary to the cynic's belief, money is not quite the universal grease for the squeaky wheel of government. Some bureaucrats can't be bought with money—they require flattery.

We do not inevitably find money as our clue to the one-to-one fallacy, but sometimes we can fall back on a rather simple line of reasoning, because we are suspicious to begin with. For example, in Figure 15.3, where we saw a typical "repressible" enzyme at work, the enzyme operated by regulating the first member of the chain of substances being synthesized. Numerous biologists have remarked that these regulators always seem to operate "early in the synthesis." From this observation, one may readily stumble into a one-to-one fallacy.

If there is a second enzyme regulating B synthesis, and if it is *also* repressed by X, then the diagram of effects would be that seen in Figure 16.9. But though *we* can see the B-enzyme in the diagram, the biochemist could easily be unaware of it in the laboratory, and thus be led to believe that the A-enzyme is the *only* regulatory loop. Why? An enzyme can only regulate the rate of synthesis if there is some precursor substance out of which to synthesize—as we have symbolized in the circles indicating multiplication of rate times quantity. If A-enzyme is repressed by X, then the amount of A will ultimately diminish, unless B-enzyme is much more strongly repressed. But once A has diminished, synthesis of B from A stops anyway—regardless of the presence of B-enzyme.

The only way to detect the presence of the second regulatory loop, involving B-enzyme, is if the first, involving A-enzyme, breaks down—just as we would be able, from black box observation, to detect the presence of the overflow pipe in Figure 16.3 only if the float-valve regulator began to fail. To experiment on such a system, we must somehow remove the one mechanism before we are permitted by the system to see the other. When we perform such an experiment, we may be surprised by the other side of the one-to-one fallacy. When we remove a mechanism, we may rudely find that it was not doing "just one thing."

Because of the separation of variables strategy, we tend to believe that each mechanism will have its unique role in the regulatory drama. But mechanisms have a way of picking up subsidiary roles, in addition to those we imagine to be—and which may well have been—their primary ones. Our lungs certainly can be said to play a primary role in the regu-

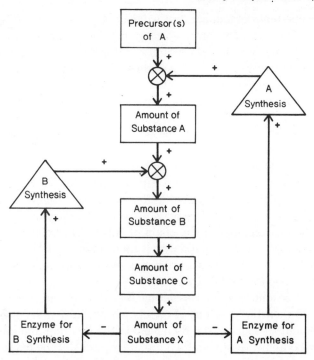

Figure 16.9. The "first-in-the-chain" fallacy would be committed here by a biologist who couldn't detect the B feedback loop unless the A-loop's repression was disabled.

lation of oxygen, but also serve as exchangers of carbon dioxide, heat, and water vapor—and, moreover, help us to float when we fall into the mill-pond. The army serves to repel the foreign foe, but also can suppress the enemy within, or build dikes in a flood, or educate the youth of a nation to a particular political philosophy, or simply act as a buffer for excess labor so as to regulate the going wage.

Or consider the question of "the function of the state." In Marxist theory the state is an instrument of oppression of one class by another—and thus will "wither away" when there are no more class differences. But we see, for example, in the Soviet Union, that the state has not yet withered away after half a century of "the dictatorship of the proletariat," which is supposed to oppress—and thereby obliterate—the bourgeois class. Although Marx and Lenin explicitly indicated that no precise predictions about the *time* for withering away could be made, their critics are bound to expect that after 60 years *some* sign should appear.

A possible explanation of the failure of the state to wither away so far

in the Soviet Union is the one-to-one fallacy. In the first place, though the
state may be an instrument of oppression of one class by another, it may
not be the *only* such instrument. If other instruments exist, then they will
prolong the life of bourgeois oppression of the proletariat. Secondly,
though the state oppresses, it doubtless has other functions to perform,
such as seeing that the people eat and the trains run on time. As long as
a complete separation of variables has not taken place, the state will not
be entirely free to wither away.

The Kool-Aid Fallacy and the Aspirin Illusion

Alice looked round her in great surprise. "Why, I do believe we've been under
this tree the whole time! Everything's just as it was!"

"Of course it is," said the Queen. "What would you have it?"

"Well, in *our* country," said Alice, still panting a little, "you'd generally get
to somewhere else—if you ran very fast for a long time as we've been doing."

"A slow sort of country!" said the Queen. "Now, *here* you see, it takes all the
running you can do to keep in the same place . . ."

> Lewis Carroll
> *Through the Looking Glass*

Like Alice, who had to run very fast to keep in the same place, the regulator, trying to hold a variable as steady as possible in the face of varying
environmental influences, must meet each variation with some compensating action. This is Cannon's Perversity Principle: the regulator must
behave in a manner contrary to the desired behavior of the regulated variable: the regulator must be active so that other parts may be passive.

A common fallacy begins with the confusion between the regulator and
the thing regulated. On a hot day, we drink iced drinks, but the ice cools
the "thermostat" in the hypothalamus, and we suffer. This is the essential
cycle of the "Kool-Aid Fallacy." In the mountains, our landlord lives
downstairs, but there is only one thermostat for the two apartments. By
some idiocy of design, that thermostat is on the wall next to his fireplace,
so when it gets cold, he lights a fire and we freeze.

A construction firm created a computer system for controlling jobs.
When expenses on a job got high, however, this firm first economized by
slashing the budget for job control. They were upset to find that expenses
grew bigger: materials were lacking when needed and tasks were not finished in the proper order, so the expense for people killing time on the

job grew by leaps and bounds. Astonishingly, they could not comprehend what was taking place, for they had a formula that measured the job control function by taking the ratio of control costs (C) to total costs (T). By reducing C, they reduced the ratio of C/T in *two* ways—directly and through the increased costs on the job making T larger. Finally, they were made to see the Kool-Aid Fallacy of it all. Regulatory costs (C) simply could not be compared directly with the costs of the variable they were trying to regulate (T). The persuasive argument was to show that the control department could make itself look twice as good by ordering materials to be dumped in the ocean on the way to the construction site.

In medicine, the Kool-Aid Fallacy frequently appears in the guise of the Aspirin Illusion—the suppression of pain instead of the eradication of the disease for which the pain is a warning. To be sure, it's sometimes necessary to suppress pain in order to keep the pain itself from destroying the system, as by driving a person to suicide. A fire alarm must be loud enough to be certain of being heard, but not so loud as to deafen the firefighters. Better the house should burn down.

The Aspirin Illusion is the deadly form of the Kool-Aid Fallacy. It is perpetrated by the general who court-martials the captain who brings the information that the enemy is winning the battle. So, too, the manager who fires the programmer who says the new computer system cannot be made to work within the schedule and budget. And though the fallacy is immediately deadly for the captain and the programmer, it is ultimately deadly for the general and the manager, too. Once you've murdered a few messengers, where will you get the news you need to lead the troops?

The Kool-Aid Fallacy also appears in medicine in less deadly shrouds. In the political furor over cyclamate sweeteners, the main argument for their continued use was that they served an important medical function—for diabetics and others needing to control sugar intake. Yet as cyclamate usage rose to the sweet equivalent of 600,000,000 pounds of sucrose, no decrease in the consumption of ordinary sugar was observed. Could it be that the entire medical argument for the use of sugar substitutes is fallacious?

Several reports have indicated that sweetness may serve a regulatory role. Sweetness in the mouth may stimulate regulatory activity to lower blood sugar in anticipation of the arrival of sugar in the digestive system. If the sweetener is artificial, the system's model fails and no sugar arrives, leaving the blood sugar at a depressed level. But depressed blood sugar stimulates appetite, so the net result of using artificial sweets could be to complicate the already difficult life of the obese. Perhaps it would be best to take your Kool-Aid with real sugar—and warm.

The False-Alarm Fallacy

"I was wondering what the mouse-trap was for," said Alice. "It isn't very likely there would be any mice on the horse's back."

"Not very likely, perhaps," said the Knight; "but, if they *do* come, I don't choose to have them running all about."

> Lewis Carroll
> *Through the Looking Glass*

The Kool-Aid Fallacy takes its name from physiology, but it spans the entire range of human foibles. The overworked library staff takes measures to discourage the circulation of books. The police department suffers criticism because there aren't enough arrests. When nobody gets smallpox, vaccination is labeled a waste of the taxpayers' money. And when the war beings to dwindle into a firefight, the antiwar demonstrators go home discouraged.

Police, public health, and political demonstrations all illustrate another important fallacy, which we shall name the False-Alarm Fallacy. The False-Alarm Fallacy occurs because regulator and regulated are in a complementary relationship. As the regulator succeeds, its very purpose may be forgotten. Worse, as Vickers observed, we can't know for certain which parts of the regulator's activity have been successful—or if indeed the success was even related to the regulator.

As the regulator falls into disuse by virtue of success, it may wither from lack of exercise. If mice *do* come, it's unlikely, knowing the White Knight, that the trap will still work, for it won't have been tested for years.

To destroy the antiwar movement, a president can temporarily diminish warlike activities. A new aggression can be started in a day, but the antiwar movement will require months to regain its vigor. Of course, a president mustn't wait too long between skirmishes, lest the army itself fall victim to the False-Alarm Fallacy. The prudent policy seems to be intermittent wars, often enough to keep the army in practice, but not often enough to meet serious resistance on the home front.

When the mechanism's *physical* basis atrophies, the results can be spectacular. The grain reserved against years of drought may rot in the bins if there is a long succession of bountiful years. But the informational parts of the regulatory mechanism also atrophy when not used, and because they are less conspicuous, the decay is likely to proceed further before anyone notices. A businessman is not likely to forget one of his bank accounts, but he quite likely *never* thinks about the information stored in his files or in the heads of his employees—information which is just as vital to the regulation of his affairs.

At times, the atrophy of information can be spectacular, particularly when the information to be stored grows so massive that its sheer physical size becomes a problem. Some years ago, we worked with an accounting firm that kept records for small oil producers. Over the years, the firm had accumulated over 25 million punch cards on which were recorded the production records of most of the wells in the area. Such information becomes extremely valuable when an oil pool begins to be depleted and measures are to be taken to extract as much of the remaining oil as possible. But when the accountants went to retrieve the cards for just that purpose, they found that half of them had been eaten into by rats and bugs. The information was utterly destroyed. Millions of dollars worth of oil may never be recovered from that pool, all because the firm never recognized that information must be stored in some physical medium and thus is subject to deterioration.

We must not imagine, however, that atrophy of information comes only through the direct deterioration of the physical medium in which it is stored. Information can become unusable over time in a vast number of ways: new codes are used and the old ones are forgotten, files are rearranged and critical items become impossible to find, new ways of calculating derived quantities come into use and nobody realizes that old and new figures are not really comparable. In many cases, the deterioration takes place on the information that is in hidden storage in the minds of the people who work with it. Either they forget things or, over a period of time, all the people who might remember how to read the old files have left the organization. The language of the files becomes another undecipherable lost language, like Etruscan or the Mayan glyphs, awaiting its own Rosetta Stone and a Champollion to break the code.

Systems designed to regulate against the possibility of extremely rare events must be well regulated themselves against atrophy. Schools have fire drills; peacetime armies have maneuvers—or statesmen encourage "small" wars to keep the army in practice. It is for this reason that we carry many of our "primitive" reaction systems—like the alarm reaction—so long after they have ceased to be functional, or have even become dysfunctional in modern life. But it is in human organizations that the need to regulate the rarely used regulatory mechanism is most evident.

In a certain midwestern city, a new city hall was built combining all the functions of government into the most modern physical plant. Among the technological wonders in this city hall was a system of alarms connected to every bank in town, operated during working hours by a police sergeant. In case of an attempted bank robbery, all the teller had to do was inconspicuously step on a signal button and the police would be alerted instantly.

Three years after this alarm system was built, it had its first trial by fire. A plain-looking young man walked into a bank near the university campus during the morning rush and handed a teller a paper bag with a note reading:

"Fill this bag with tens and twenties. I have a gun."

The teller calmly stepped on the alarm and began slowly filling the bag —but the police didn't come. Finally, when the teller could stall no longer, he handed over the bag with $8000 and the bandit melted into the crowd, never to be seen again.

The police were called immediately, but it was too late.

Somebody asked why the alarm system hadn't worked. When they went into the alarm room to investigate, they found the sergeant working in the back of the control panel with a screwdriver.

"Hey," he said, "come take a look at this stupid machine. One of the lights has gone on and I can't seem to make it go out."

While it is easy to criticize the sergeant—now a cop on the beat—the fact remains that he made a perfectly reasonable assumption. There had never been a robbery before, but there had been lots of malfunctions of the machine. If the bankers in town didn't want the system to atrophy, they should have held irregular drills—to test out the police, not the hardware.

This, then, is the False-Alarm Fallacy, the belief that certain regulatory systems remain in perfect working condition, as long as they don't change *physically*. It ignores the *role of the model*, and the need for any regulatory system to maintain an up-to-date model of its environment and of its own behavior.

Originally, we intended to conclude this section with the story of the poor sergeant, but a few days after this was written, we ran across the following news item in the local paper. Because the story completes the tale in an entirely different culture, yet still illustrates the False-Alarm Fallacy to perfection, it demonstrates how truly general a systems principle it is.

Officials tripped the alarm system at a bank in a rural area outside Bangkok in a test to see how long it would take police to arrive. The only response was a small boy carrying a bowl of noodles. Investigation showed that the police station where the alarm buzzed was located next door to a noodle shop, and the bank employees had arranged with the police to use the system to order their lunch.

Flareback

"Let's fight till six, and then have dinner," said Tweedledum.

"Very well," the other said, rather sadly; "and *she* can watch us—only you'd better not come *very* close," he added: "I generally hit every thing I can see—when I get really excited."

"And *I* hit every thing within reach," cried Tweedledum, "whether I can see it or not!"

Alice laughed, "You must hit the *trees* pretty often, I should think," she said.

Tweedledum looked round him with a satisfied smile. "I don't suppose," he said, "they'll be a tree left standing, for ever so far around, by the time we've finished!"

> Lewis Carroll
> *Through the Looking Glass*

Preventing atrophy in the regulatory mechanism is just one example of the necessity to regulate the regulator. Once the regulatory mechanism itself becomes tangled in the regulatory plait, further possibilities of confusion arise. A political system—the state, for example—is essentially a regulatory mechanism. Yet in modern states the cost of this mechanism becomes a major burden on the populace, thus tempting actions to regulate the cost of government, the regulatory mechanism itself. For various reasons, cost being only one, it is a profound truth that "he governs best who governs least," or, as the Chinese put it, "governing a large country is like boiling a small fish."

To some extent, the Kool-Aid Fallacy applies to government, but just because a government is *supposed* to be regulating does not mean that all its activities are to good purpose. There is, no doubt, in every government, some functional "waste" and some real waste. Because there is real waste, there will always be pressure to consolidate functions of government to achieve "efficiency," and this pressure may complicate the regulatory problems.

The strategy of separation of regulatory problems is nowhere more useful than in facing the problem of *protecting* regulatory information. The American Constitution prescribes the taking of a census, because the founding fathers well understood the usefulness of a good census for the regulatory task that is government. They also knew that the census would be worthless if not accurate, and not accurate if the persons surveyed had any reason to fear being accurate. Thus the census information is protected by strong law, and even stronger tradition.

But this insight was not original with the founding fathers. They probably got it more or less directly from Hume, but it was understood centuries before by Chinese scholars. Ma Tuan-Lin, a thirteenth century scholar, knew quite well that one could not get a reliable census from figures taken in the course of imposing taxes or registering men for military service, where the populace had every incentive to give incorrect information. While it would doubtless be "cheaper" to combine the census with the tax system, the regulatory efficacy of the census would probably be destroyed in the process.

The temptation to intervene *directly* in complicated regulatory plaits is understandable—but inexcusable, in the light of all our knowledge of the consequences. The habitual result is directly contrary to the direct intervention—though we never seem to learn from the habit. This contrary reaction is so ubiquitous that Garrett Hardin has given it a name:

To get rid of insects, we spray promiscuously with such potent poisons as Malathion. As a result, we kill not only millions of insects, but thousands of birds. Because birds are a great natural negative feedback for insect populations, using insecticides often causes a secondary *increase* in the number of insects later. We may refer to this as a "flareback"—thus verbally acknowledging our failure to think in terms of systems.

Unfortunately, Hardin's hasty verbal sketch may leave a fallacious impression about flareback, which we must now erase. Figure 16.10 shows an effects diagram of Hardin's verbal statement. The solid line from chemicals to insects is our intended direct intervention, while the dotted line shows the "accidental" effect because the chemical happened to kill birds as well as insects. The problem with this representation is that it seems to imply that if we are less promiscuous in our spraying—or if we choose chemicals more carefully so as not to harm birds—we can eliminate the flareback.

But flareback is a much more subtle process, and more often arises not from our clumsiness as killers but from our virtuosity. Suppose, for example, that we perfect a chemical which is *absolutely* harmless to birds, and almost completely lethal to insects. In that case, the line in Figure 16.10 from chemicals to birds has to be erased. But in Figure 16.10, we have neglected to draw another important line—that from insects to birds. Not only do the birds regulate the insect population, but the insects return the favor for the birds. Only neglect of the Principle of Indifference makes us talk or think as if we know which is regulating which. Our entanglement with "purpose" may lead us to elevate the one direction over the other, but regardless of what we think, the regulation is *mutual*.

Once we have drawn this new line, as in Figure 16.11, we can also draw the transitive closure (Figure 16.12) which gives us back the dotted

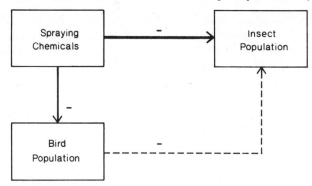

Figure 16.10. Simple flareback results when the spraying of chemicals to kill insects also kills birds, who are predators on the insect population. The reduction of the bird population allows an increase in the insect population, just the opposite of the intended effect of spraying.

line from chemicals to birds—but *by way of* insects. From a regulatory point of view, it matters little whether the birds are poisoned or starved—though poisoning may make *us* feel more guilty. The net effect is a drastic reduction of the bird population. This might be all right—who likes all those noisy birds in the morning anyway—except that there is almost no chance that we will succeed in poisoning *every* last insect.

Insects can multiply much, much faster than birds, so *their* population will flare back even if not every bird has starved to death, and it will be a long time, if ever, before we return to the happy state where there were only a moderate number of insects.

Fundamentally, flareback occurs because among the strands of the plait of regulatory mechanisms there lurk polar pairs, like Tweedledum

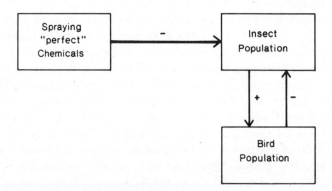

Figure 16.11. The simple view of flareback ignores the effect on the bird population of a sudden reduction in insect population caused by a "perfect" insecticide.

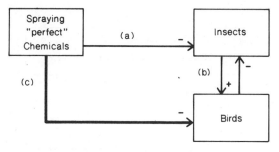

1. Secondary Effect

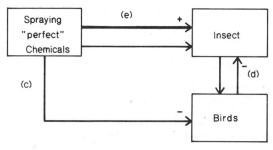

2. Tertiary Effect

Figure 16.12. The perfect insecticide has a secondary effect (a and b produce c) on the bird population. Through this secondary effect, the insecticide has a tertiary effect (c and d produce e) on the insect population. In the end, there are more insects than originally—demonstrating the *intrinsic* nature of the flareback, independent of the insecticide.

and Tweedledee, ready to do battle. When both members of the polar pair are present, each acts as a restraint on the other, but once Tweedledee is removed, Tweedledum will certainly begin to hit the trees. In Figure 16.12, we may fail to notice Tweedledum, for he is hidden in the guise of the line from insects to insects which we have omitted by convention. In all biological populations, that line is present—the "power of population," which is at once the gift of God and the Devil's curse.

But it is not population, but *polarity*, which is the ultimate source of flareback, as we can see in cases where there is no population or where the attempted intervention is in the direction of increase, thus cooperating with the tendency of the population to enhance its own growth.

The example of the wolves, the deer, and the forest (Figure 16.13) is particularly distressing, because it has happened so often in so many different places. It starts when some naive nature lover decides that

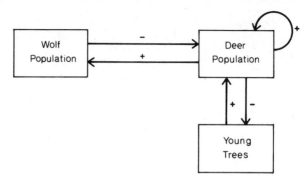

Figure 16.13. The wolf-deer-tree system demonstrates that flareback has nothing to do with eradication strategies, for here attempts to directly *enhance* the deer population ultimately lead to eradication of wolves, deer, and even the trees.

wolves are bad because they eat deer. A bounty is put on wolves, and the wolf population is suddenly reduced. Sure enough, the deer population increases—so much so, in fact, that long before the winter is over, the deer have eaten all the available food. All the deer starve to death, the rest of the wolves die—after a last supper of starving deer—and the forest is wiped out because all the young trees have been eaten bare. Even this analysis is magnificently oversimplified, for the ecological system contains vastly more parts than wolves, trees, and deer, and vastly more than the simple interconnections that first meet our eye.

In systems of any complexity at all, we know that almost every part will be involved in a mulitude of regulatory mechanisms. Therefore, when we hear "I just want to change *one* thing," we mobilize for trouble. Consider, for example, the type of enthusiasm which greets each new technique of insect control. The gypsy moth, for instance, is a "pest" in many forest areas, and some years ago, scientists isolated the sex attractant of the gypsy moth, first called gyplure, presumably for "gypsy lure." The idea, of course, was to spread this material around forests infested with gypsy moths, in order to confuse their mating behavior. The male moths would be attracted to the lures rather than to the females. Although this idea has a great deal of merit, the early responses to the discovery were far more enthusiastic than a reasonable acquaintance with the flareback fallacy and the Count-to-Three Principle would permit.

At that time, we frequently encountered enthusiastic letters such as one in *Science* which concluded:

The broadcast application of gyplure is an approach which can't conceivably do any harm, would be much less expensive than insecticides, and might possibly afford the means of finally eradicating the gypsy moth.

In the light of what we know about flareback, would it be impertinent to suggest that "can't conceivably do any harm" is not a statement about the harmlessness of gyplure, but about the stunted conceptual abilities of the writer?

Taking Complexity by the Horns

"When *I* use a word," Humpty Dumpty said, in rather a scornful tone, "it means just what I choose it to mean—neither more nor less."

"The question is," said Alice, "whether you *can* make words mean so many different things."

"The question is," said Humpty Dumpty, "which is to be master—that's all."

Lewis Carroll
Through the Looking Glass

Every time we think we've finally captured the full subtlety of some system of regulation, something new turns up. No wonder many writers and thinkers have turned, in later life, to the Humpty Dumpty Method—defining complexity in such a way that it does exactly what you want it to do.

In the Humpty Dumpty Method, you take the complexity bull by the horns by

1. Stating that "complexity" is a measurable property of a system (rather than a relationship between system and observer, as we know it to be),
2. Stating, or "proving," that this-and-such property (such as "stability") follows from high (or low) complexity.

In this way, Complexity, in all its complexity, is brought under control by the primitive device known to anthropologists as "name magic"—if I give it a name, I control it.

Margalef (1963) was the first of the modern ecologists to "demonstrate"—through the use of information theory—that systems with more different species in interaction were consequently more stable than less diverse systems. The greater number of interactions was supposed to account for this increased stability, though Ashby (1952) had previously shown on much more general grounds that there was no necessary connection between increased diversity and either increased or decreased stability:

. . . in adapting systems, there are occasions when an increase in the amount of communication can be harmful. (Ashby 1952)

Because Ashby was a psychiatrist, engineer, and general systems thinker, his work seems to have been ignored by ecologists. This is to be expected in a new field where territorial disputes within the field allow little time for exploration outside the newly established boundaries. Thus Margalef's diversity-stability hypothesis stood for a few years as "intuitive" to the majority of ecologists, professional and amateur.

Eventually it came under attack from all sides. May (1973) did mathematical modeling that seemed to show, if anything, the opposite—that more complex systems were less stable than simpler ones. Closer empirical observation turned up many cases in which the old, "mature," diverse ecosystems were very sensitive to the slightest disturbances. And not all the "mature" systems turned out to be more diverse in the first place.

By now, the diversity-stability hypothesis seems to have run its course as a scientific fad, but our interest is in why it became a fad in the first place. Quite likely, the major reason is expressed by Ehrenfeld:

In our eagerness to demonstrate a "value" for the magnificent, mature, and diverse ecosystems of the world—the tropical rain and cloud forests, the coral reefs, the temperate zone deserts, etc.—we stressed the role they were playing in immediate stabilization of their environments. (1976:652)

Supporting Ehrenfeld's analysis is the frequency with which this same kind of "eagerness to demonstrate a 'value'" had led to political arguments cloaked in the rhetoric of "stability." One much older example, from sociology rather than ecology, will make the point clear to modern readers free of the male supremist notions of the nineteenth century. Emile Durkheim wrote in his classic work, *Suicide,* that

. . . woman can endure life in isolation more readily than man . . . it is because her sensibility is rudimentary rather than highly developed. . . . With a few devotional practices and some animals to care for, the old unmarried woman's life is full. If she remains faithfully attached to religious traditions and thus finds ready protection against suicide, it is because these very simple social forms satisfy all her needs. Man, on the contrary, is hard beset in this respect. As his thought and activity develop, they increasingly overflow these antiquated forms. But then he needs others. Because he is a more complex social being, he can maintain his equilibrium only by finding more points of support outside himself, and it is because his moral balance depends on a larger number of conditions that it is more easily disturbed. (1951:215–216)

This argument must have been terrifically convincing to Durkheim's contemporaries, half of them male and hard pressed to reconcile the larger male suicide rate with their presumed superiority. Durkheim managed to twist it around so that the higher suicide rate was *proof* of superiority. Complexity, once again, was better—but isn't suicide *less* stable? So in this case, *simpler* was more stable.

If you search the old literature in almost any social or biological science, you'll find these arguments about the relationship between simplicity, or complexity, and stability. You may even, by misreading this chapter, think that we're trying to take a side in this argument. After all, didn't we go to some length to show that a better toilet could be built by adding various regulatory mechanisms to its plait?

If you thought that, please read the chapter again, for you have fallen into the clutches of simple-minded complexity. What we demonstrated was that a *different* configuration of mechanisms led to different performance in a variety of environments, which could be better or worse, depending on the other dimensions. What we've tried to do is counteract the majority tendency to oversimplify the behavior of regulating systems, but in doing so, we don't want to appear to advocate oversimplifying complexity itself.

Let's consider the toilet problem in the light of these definitions of "complexity." In the first place, because we know that complexity may simply be in the eye of the beholder, it can hardly be the basis for universal laws—unless these laws are laws about the minds of beholders, as well as about the systems beheld. This is the approach taken by Ashby in his classics, *Design for a Brain* (1952) and *Introduction to Cybernetics* (1956), which all thinkers on this subject should consult.

Secondly, assuming the beholder problem has been put aside or encompassed within the meaning of complexity, what does happen to the stability of the toilet system as we add one mechanism after another to take care of ever less probable threats, like the White Knight? Adding a mechanism is a regulatory strategy. Like all strategies, it comes, eventually, to *some* limiting factor. Mechanisms carry a cost, both to build and to maintain. Eventually, the cost of building or maintaining a particular mechanism becomes larger than the value to the system. At that point, additional complexity *reduces* stability.

A good way to see this limitation is by considering the complementary strategy—the purchase of redundant toilets. When the cost of gadgets gets big enough, the price of one complex toilet will be greater than the cost of two simple ones, but will the one perform better, in the survival sense, than the two? And, if not two, then what about the toilet so full of gadgets that it costs as much as 100 simple ones? Perhaps there is a message in the propensity of the newly prosperous to buy houses with *many* toilets, rather than with one *complex* toilet?

But then, we'd better not go too far in the other direction, either, for the newly rich won't generally buy a dozen outhouses in place of one flush toilet. If there's any word that's even more complex than "complex," it's "simple." After all, if we could define a unique "simplicity" measure, we could invert it to obtain a measure of "complexity."

Perhaps it would be wise to leave the last words on this subject to Ashby. In speaking of the brain, he manages to encompass both the problem of the observer and the problem of stability in a single package, so that his sensible approach to these subjects has not yet been surpassed:

Doubtless there are even more factors to be reckoned in the balance, but what we have seen is sufficient to show that *richness of connexion between the parts of the brain has both advantages and disadvantages. . . .* Thus, for the organism to adapt with some efficiency against the terrestrial environment, it is necessary that the degree of connexion between the reacting parts lie between certain limits. (Ashby 1952)

This is not only an excellent statement of the kind of regulatory strategy we are most likely to find in systems, but also of the way we ought to approach the problem of understanding those systems—not too simple, but not too complex, either. Instead of taking complexity by the horns, we simply escape between them—not as satisfying, perhaps, but far safer.

QUESTIONS FOR FURTHER RESEARCH

1. *Population Policy*

Malthus, in his *Essay on Population* (1964), pointed out that there were two general sorts of checks on population, which he called "positive" and "preventive." Positive checks were those that removed persons by death, whereas preventive checks were those that reduced the entry of people by birth. Malthus recognized numerous positive checks—war, pestilence, famine—but, perhaps because he was a parson, he was rather limited in his perception of preventive methods. For the most part, though not denying other possibilities might exist, he saw prevention in terms of postponement of marriage. List as many positive and preventive checks on population as you can. Discuss the "costs" of each method, and how the methods interact. Finally, discuss how the existence of multiple mechanisms complicates the job of planned population regulation from above.

2. *Sociology*

List the regulatory functions that slums provide for a society and discuss what implications this list has for slum clearance projects, or for less drastic means of removing slums.

3. *Social Welfare*

Discuss the following two campaign promises:

1. "If I am elected, I will reduce the number of people on welfare by forcing all the able-bodied to find jobs."
2. "If I am elected, I will spread welfare benefits to many people now not covered."

In particular, what flareback effects might be expected when an attempt is made to push the variable, "number of people on welfare" *directly* in one way or the other.

4. *Good Luck*

There is a story—the truth of which we cannot verify or refute—which says that people who had won prizes of $50,000 or more in the Irish Sweepstakes were found to have a significantly increased suicide rate in the succeeding 10-year period. Can you present a plausible model of how this effect might have come about? What strategies might one adopt, upon winning a sweepstakes prize, to prevent such a mournful outcome?

5. *Drug Addiction*

Discuss the following idea:

. . . the surest and simplest, if not the quickest way for Britain to get a drug addiction problem similar in proportion to that of the United States would be to try to reduce the problem it now has by any repressive means, or indeed by any means which gave the opportunity for the "image" of drugs, drug use or addiction to change so as to generate a deviation-amplifying system. (Wilkins 1968:426)

6. *Sociobiology or Biosociology?*

The "first-in-the-chain" fallacy need not be confined to biochemistry (see Figure 16.9). One of the raging controversies of our time is whether biological or cultural factors "truly" regulate the behavior of people. Construct an example with the structure of Figure 16.9's nested feedback loops in which

1. The outer loop is biological and the inner, cultural.
2. The outer loop is cultural and the inner, biological.

Write a script for an argument between a biologist and an anthropologist over each of these two cases. (References: Wilson 1975; Sahlins 1976)

7. *Disasters*

One of the most dangerous guises in which we find the one-to-one fal-

318 Overly Simple Views of Regulation

lacy is the belief that protection by one regulatory mechanism obviates the need for others. Examples include:

1. People hear that a hurricane is being seeded to reduce its force, and so refuse to obey warnings to evacuate or otherwise protect themselves.
2. A diplomat believes that her country's defense against missiles is impregnable, so she neglects to display the customary diplomatic amenities to representatives of other nations.
3. A motorcyclist, forced to wear a helmet, imagines that accidents are no longer a source of concern, and so drives 20 percent faster.

Give three examples of this "bulletproof vest" fallacy. Describe how people could be influenced to protect themselves even when they are being convinced by other sources that they are already protected sufficiently to allow foolish behavior. Then give three examples of the opposite fallacy, such as a woman who continues to take birth control pills after a hysterectomy.

8. *Plumbing*

The overflow pipe in Figure 16.2 protects against flooding if the valve on the input line leaks. If the same valve should happen to stick in a closed position, the tank will not refill, so the toilet will no longer operate. Show how another mechanism could be added to back up the float valve system when, for some reason, its input valve is shut off. Describe the regions of the environment's state space in which this new mechanism will prove useful. Contrast the relationship of this redundant mechanism to the float valve with the relationship that the overflow pipe has to the float valve.

9. *Evolution*

Give a plausible explanation of how nested regulatory feedback loops, such as in Figure 16.9, could have evolved over time. Try to find an example, not necessarily in biochemistry, in which such an evolution is taking place, but not quite completed, so that both feedback loops can be seen or experienced at the same time, at least in some situations.

17

Blindness and Reversed Vision

"Now, if you'll only attend, Kitty, and not talk so much, I'll tell you all my ideas about Looking-glass House. First, there's the room you can see through the glass—that's just the same as our drawing room, only the things go the other way. I can see all of it when I get upon a chair—all but the bit just behind the fireplace. Oh! I do so wish I could see *that* bit! I want so much to know whether they've a fire in the winter: you never *can* tell, you know, unless the fire smokes, and then smoke comes up in that room too—but that may be only a pretence, just to make it look as if they had a fire. Well, then, the books are something like our books, only the words go the wrong way: I know *that,* because I've held up one of our books to the glass, and then they hold up one in the other room."

Lewis Carroll
Through the Looking Glass

Adults know that the "other side of the mirror" is not real, and this adult wisdom is carried over into systems thinking in the form of ignoring or misinterpreting the role of the environment—the other side of the system mirror. Probably the most common way of ignoring the environment is through violations of the Piddling Principle, but this is only a special case of the dependence of system on environment for regulation. What every regulatory system ultimately needs from the environment is *constraint,* in one of three ways:

1. As a direct *source* of regulation, as the Piddling Principle indicates.
2. As a *reference* which can be used as a set point for regulation.
3. As the *basis* for building a regulatory model, as indicated by the Parallel Principle.

Each form of dependence has its own associated fallacies.

But the final fallacy is suggested by Alice when she tells Kitty how she *knows* about books in the Looking-glass House. She knows because she *sees*. Whatever else happens, we believe our own eyes. Yet it's a tiny step over the precipice from "I know it exists because I see it" to "I know it doesn't exist because I don't see it."

Hidden Reserves

> "You are old," said the youth, "and your jaws are too weak
>   For anything tougher than suet;
> Yet you finished the goose, with the bones and the beak—
>   Pray, how did you manage to do it?"
>
> "In my youth," said his father, "I took to the law,
>   And argued each case with my wife,
> And the muscular strength, which it gave to my jaw,
>   Has lasted the rest of my life."

>   Lewis Carroll
>   *Alice in Wonderland*

Father William is a rarity among us old codgers, for he remembers accurately the humble origins of his numerous talents. Carroll's Father William is a parody of Robert Southey's original Father William—one who was more like the rest of us. He was inclined to answer the honest questions of a young man with pious maunderings about his excessively moral youth—answers that make youth grow up thinking that balance, agility, strength, and a keen eye have something to do with religious faith.

Bombarded as we are with falsified personal histories, or national histories, we take our strengths for granted. Something has always saved us in the past, so we needn't think where it came from. The football team is on the brink of defeat, but somewhere finds the "inner strength" to hold that line four more times. Inner strength? Perhaps. But perhaps it has been built up from a training program, much like Father William's, stretching far into the past, much before the game, and even before the players matriculated.

The office staff is working under high pressure to get out the annual report. The office manager, observing the enhanced rate at which the staff works under pressure, decides that the only thing keeping them from consistently operating at this high level is the lack of pressure. When the reports are out, she creates an artificial emergency, and sure enough, performance stays high.

Soon, however, the pressure needed to maintain performance has to be increased. Eventually, the pressure has mounted to totalitarian proportions, and the supervisor is astonished and chagrined when the entire office collapses, exhausted.

When we are busy, we may skip a meal. Afterward, when we reflect on the situation, we observe that skipping one meal didn't seem to impair our efficiency. Sometimes, we feel that it made our performance even better.

But though our body allows us to skip one or more consecutive meals without loss of efficiency, we don't imagine that the process could be extended indefinitely. The fluids and foods of the body are constantly being consumed. If they are not replenished, vital functions begin to fail, and the effect on the system is critical. A reserve allows the system to smooth out fluctuations in the environment, but you can only operate so long on reserve. Then you must pay.

When the "hidden reserve" is not in the system, but in the environment, we have the fallacy of hidden input. We sometimes hear that life violates the laws of thermodynamics. Doesn't the history of life show a constant increase in order, structure, and complexity?

Perhaps the history of life does show such a progression, but any progress depends on the continuing radiation of vast quantities of energy onto the earth by the sun—an input which has been so steady and dependable for billions of years that it is taken for granted by all forms of life. Shut off the sun and life stops.

The effect of cutting off the reserves of energy is dramatic, but other hidden reserves are equally important. Living systems of all kinds, and many mechanical systems, too, depend on the input of various kinds of structure, and failing to take note of such input can engender the same sort of illusion.

Consider the common business question of where to locate a new factory, or whether to move the old one. A major consideration in such moves is the going cost of labor in the contemplated locales. If the going wage is lower, the move seems more attractive, but a factory never relies merely on "raw" labor. To the extent that the workers need some skills, there must be training of the workers. This training cannot be detected by weighing the workers, or measuring their body temperature, but it is "in" them just as energy is in them. If it isn't "in" them, then they must get it from somewhere, otherwise they cannot operate the factory, or perhaps cannot operate it efficiently.

A company has the choice of obtaining this training by hiring workers who were previously trained—at appropriately higher wages—or of providing this training to untrained workers, at a certain training cost, or of operating the factory at lower efficiency while the new workers learn on the job. Consequently, some factories move to low-wage areas only to find that their expenses actually increase. Their hidden input of skill—which was why they were paying higher wages in the old location—may not be available in the new location—which is why wages are lower. In the old location, the payment for training took a less conspicuous form, but the level of skill is being regulated in either case, or the factory cannot operate.

It would be lovely for us if we could confidently assert that there will *always* be hidden reserves, for that would be a most powerful heuristic for avoiding fallacies and making discoveries. Unfortunately, thinking in this area is not that simple, and imagining that it *is* that simple is the opposite side of the same fallacy. How pleasant it would be if we could all believe the flowery optimism of these 1867 sentiments which still echo in many halls today:

Our mother earth holds within her bosom all the various materials needed for the preservation and well being of her children. When the woodsman's axe ruthlessly stripped her of her rich vestments of umbrageous forests, and thus awakened apprehensions as regards the supply of materials needed to furnish household warmth, we were directed to the outcroppings of black carbon in our immense coal fields; and when the Nantucket and New Bedford whalemen returned to their wharves, with the alarming announcement of the partial or complete failure of the ocean harvests of oil, the little rivulets of petroleum which oozed from the rocks of Pennsylvania were sounded to their depths, and immediately the oil spouted up in such quantities as taxed all our energies to secure. (Nichols 1967:45)

Of course, what Mr. Nichols, and others of his ilk, failed to note was that if the coal or the oil had *not* turned up in time, there would have been no Mr. Nichols writing this piece of fluff.

The Environment as Reference Point

"Are you content now?" said the Caterpillar.

"Well, I should like to be a *little* larger, Sir, if you wouldn't mind," said Alice; "three inches is such a wretched height to be."

"It is a very good height indeed!" said the Caterpillar angrily, rearing itself upright as it spoke (it was exactly three inches high).

"But I'm not used to it!" pleaded poor Alice in a piteous tone.

Lewis Carroll
Alice in Wonderland

When a system uses its environment for *reference*—rather than simply as a source of materials, energy, or structure—the relationship may be so subtle as to tempt us into easy errors of reasoning. In the first place, environmental variables which are used as set-point references are quite likely to have been *so* constant that we are unaware of their presence. Otherwise, it would have been "foolish" of the system to depend on them for reference. That's why Alice is so wretched—not because she's three inches high, but because she has a multitude of dependencies on an environment as seen from a very different vantage point.

The size of familiar objects is a reference we use incessantly for orientation. That's why buildings in the neighborhood we left as a youth seem smaller when we return as adults. Perhaps it is the same for ice cream cones and candy bars, or perhaps they really *are* smaller. How can we remember correctly? Indeed, when we dream of giants, we are said to be dreaming of childhood, when we were very much oriented to a world of giants, and where, perhaps, mushrooms and caterpillars were more important to us.

Although we change in size, the horizon stays constant for us as long as we remain on earth. The system that controls our upright position uses unconscious visual feedback from this constant horizon, because it is constant. If the horizon were not constant, we wouldn't have evolved such a system of orientation, but because the horizon is constant, we're not aware of the system.

We are reminded of the system when this constancy is violated, as when we are at sea, or when we stand upright and blindfolded for a long time. Psychologists and carnival owners remind us even more strongly by constructing trick rooms in which all visual clues about the horizon and relative sizes have been distorted. In such an environment, our constant references are lost.

Yet because we take these references so much for granted, we do not experience the room as distorted. Instead, we see balls rolling uphill or pendulums held to the side by "mysterious" forces—a safe enough sort of anarchy for which we gladly part with our dollar for admission. The horizon is so precious to us that we gladly give up the laws of physics to preserve it.

Entire cultures may come to depend on some constant aspect of their environment in precisely the same way. In that case, the analogy with entering the dizzy room is making contact with another culture. The most striking case of this kind is that of Ishi, the last survivor of the Yahi Indians of California. Ishi was rescued from starvation and introduced to American culture by the anthropologist Kroeber. The story of his reactions is compelling in many ways, not the least of which is the "dizziness" he experienced from so many new cultural reference points.

To take just one example, in Ishi's native society, "many" people had meant, at most, 40 or 50. His daily life, however, had included only about a dozen others and, for many years, only four. Given this frame of reference, it is not surprising that

No dream, no wildest nightmare, prefigured for Ishi a city crowd, its clamor, its endless hurrying past to be endlessly replaced by others of its kind, face indistinguishable from face. It was like a spring salmon run, one fish leaping sightlessly beyond or over another . . . (Kroeber 1961:133)

A hunting-and-gathering band that enters into relations with the immense productive power of the modern industrial state may find old reference standards skewed in a similar way. A material that was always in short supply may now become abundant, and since trade relations can easily be keyed to a reference of some constant shortage, the effect can be startlingly disproportionate to any "value" of the material introduced. The same kind of effect could be introduced into our own economic system by a sudden discovery of enormous quantities of gold—not particularly useful in itself, but for its reference property of scarcity. Some writers have suggested that the influx of American gold may have given the impetus to the Industrial Revolution—certainly a dizzying phenomenon equal to water flowing uphill in the eyes of the preindustrial world.

Another fine example of regulation by environmental reference is the school system. Typically, the good school shapes the environment for the student so as to maximize individual learning. Care is taken to see that reasonable quiet is maintained, that all necessary materials are at hand, and that punishments and rewards (grades) follow directly from the acts they are intended to suppress or reinforce. Unfortunately, the product of 15 years in such a well-regulated environment is a person who cannot learn in any *other* environment. When there are no more pencils, no more books, and no more teachers' dirty looks, there is also no more learning. Educators usually take this postgraduate inertia as proof that without the school there would have been no learning at all. Only a few of the finest schools—and they are not the ones that get the best ratings—know enough to make themselves unnecessary to the student's further progress. Perhaps we teach best by teaching least.

The careful regulation of the classroom environment may actually have a deleterious effect on learning, because there are different kinds of constancy to which a system may be adapted, not just the following kind:

Since the first "how to study" pamphlets shepherded students into well-lit alcoves sealed from sound, educators have recommended sterile surroundings as the most fruitful for academic work. Anything that shielded the student from distraction was a study aid. Lecturers went out of their way to present serious materials demanding more than ordinary concentration in halls antiseptic to noise. (Keating and Brock)

Keating and Brock, who start their study of "distraction" with these words, proceed to demonstrate what many of us have always suspected— that a creature adapted to a noisy environment might find silence more distracting than noise.

The same observation has been made in other contexts. Any author knows that without occasional jarring, the readers are more likely to close their eyes than open their minds—why else would this book be so full of absurdities? The essence of this observation is that *constant fluctuation is a kind of constancy.* A system may thus be adapted to using constant fluctuation as an integral part of its regulatory mechanisms, so that if fluctuation is suppressed, the system may go haywire.

Weinberg (1965) has demonstrated how the bizarre take-off of the Kwakiutl Indians' potlatch could be traced to the removal of fluctuation in the food supply that occurred with their increasing integration with Canadian culture. If you think this is an absurd conclusion, just imagine what would happen if the stock market ceased fluctuating.

Relating back to the school problem, we see that the idea of constant fluctuation puts a limit on what "progressive" schools can do to counteract the "perfect and unchanging environment" problem in the ordinary school. As one little girl in a progressive school put it to her teacher, "Do we *have* to do whatever we want again today?" Quite likely, her name was Alice.

The Homunculus Fallacies

"The Eighth Square at last!" she cried as she bounded across and threw herself down to rest on a lawn as soft as moss, with little flower-beds dotted about it here and there. "Oh, how glad I am to get here! And what *is* this on my head?" she exclaimed in a tone of dismay, as she put her hands up to something very heavy, that fitted tight all round her head.

"But how *can* it have got there without my knowing it?" she said to herself, as she lifted it off, and set it on her lap to make out what it could possibly be.

It was a golden crown.

Lewis Carroll
Through the Looking Glass

How, indeed, if Alice, the pawn, had not carried within her, for the entire journey through the Looking Glass World, the hidden potential of every one of us pawns, to become a queen? But the pawn doesn't *look* like the queen, and we saw nothing on Alice's journey that would make her look like a queen, though there were many strange things on the way. There *must* be a clue in one of them, telling us how this mysterious transformation came to pass.

Because we know that regulation responds to constraint in the environment, we do depend—through application of the Parallel Principle—on the environment for clues about regulation. In reading these clues, we are reading, as Alice did, in a mirror. Sometimes "the words go the wrong way," which is our Perversity Principle, but there are other problems associated with searching the environment for the source of the system's model, or with examining the system for signs of the environment.

Perhaps the most absurd, yet oldest, of these fallacies is the thought that the system's model must somehow *look like* the environment, which is related to the thought that for a pawn to be converted to a queen on reaching the Eighth Square, it must somehow have *looked like* a queen before. The similarity of a pawn to a queen—or to a knight, rook, or bishop, to which it may also be converted—is an abstract one, governed by the rules of chess, which are, of course, arbitrary. In the same way, requiring that the regulator's model of the environment must somehow look like the environment is about as absurd as requiring an "intelligent" machine to scratch its scalp with a pencil while it is "thinking."

This fallacy has been with us at least since the problem of Plato's Cartylus: "How does the sound of a word proceed from its meaning?" Humpty Dumpty believed that words *should* resemble the things they represent:

. . . *my* name means the shape I am—and a good handsome shape it is, too. With a name like yours, you might be any shape, almost.

but then he lived in the Looking-Glass World.

Medieval scholars believed that the embryo must look like a little person—the homunculus. In their honor, we can call this particular mental problem the *Homunculus Fallacy*—that the system's model must bear a *physical* resemblance to the environment. It's certainly *possible* that a system could literally carry a *picture* of its environmental threats. The Postal Service does that when it publishes photographs of criminals wanted for postal fraud and other unspeakable acts. Possible, yes. Necessary, not at all.

When we are climbing in the mountains, we may have been told "never reach barehanded up to a ledge whose top you cannot see!" We follow this injunction, never realizing quite why, and we are safe from rattle-snake bites even if we've never seen a rattlesnake, or a picture of one, or even heard the word. What does our "model" look like of this part of the environment? It could be anything, including some completely erroneous images, like the thought that the injunction has to do with the possibility of sharp rocks on the ledge.

Advertisers know quite well that the model need have nothing to do physically with the regulatory threat or threats it supposedly guards against. Millions of housekeepers prance gaily around the kitchen, bath, and closets spraying "disinfectant" against "germs." Ask one of them to draw a picture of a "germ," and if you get anything, it will be something conjured up on Madison Avenue to generate maximum repugnance—and thus maximum sales of the magic spray. Anyone who can successfully create a new "threat" in the minds of the masses can become rich selling scented water—because the model need not represent a threat at all, let alone resemble one.

Even the most sophisticated scientists are not immune to the Homunculus Fallacy in one form or another. We form an image of what we are seeking, and this image blinds us to the thing we're really after when it's right in front of us. When the neurophysiologists studied Einstein's brain after he died, they didn't find any fourth-dimensional space at all, but there was something there that seemed real enough to him. And in the "germ plasm" itself, where we would expect to find the original homunculus, generation after generation of microscopist has "seen" ever more sophisticated homunculi.

In social sciences, we often search for the "actor" in a group—the homunculus whose individual character or personality determines the character or personality of the group. And, just often enough to throw us off course, there *is* a charismatic person who can truly be said to be the model for an entire group of followers. But once again, the *model* of the environment need not be literal, so just because we can ascribe individual human characteristics to a group of people, there's no reason to believe we're going to find a person like that somewhere inside.

Remembering these things about models may keep us out of some trouble when searching for the regulatory model, but we'll be in even deeper trouble when there is no particular model at all. The model, after all, is built up from information fed back from the past. A system *without* a past may not have *any* model. For the Parallel Principle to apply, the system must have a history.

An animal species may have behavioral mechanisms that help regulate its population—mechanisms that are polar to the explosive potential of self-reproduction. One such mechanism is territoriality, in which, say, a male defends "his property" against intruding males. If males driven to the fringe of the habitat cannot find mates, this territorial behavior ensures that the habitat will not become overpopulated.

But, certain authors to the contrary notwithstanding, territoriality is not universal. Some populations subsist in habitats so unfavorable that the environment itself provides all needed limitation. In such an environment, there is no reason to believe that a population will acquire self-regulatory mechanisms, such as territoriality—for "territoriality" is a model of a genial habitat of the past. If a population never gets an opportunity for self-regulation, it will never learn how.

The Parallel Principle says that a system will be like its *past* environments, so we may be confused when we see a system in a present environment that has, unbeknownst to us, recently changed. We act on the inputs of middle age with the models formed in our youth—no wonder we look so fat and foolish. But the folly of youth is the folly of having no past at all—in which case we could hardly expect to find *any* models, let alone models of proper etiquette. After the revolution, new models must be made, which may be why we have such trouble understanding revolutions.

Denying the Existence of Regulation

"You know very well you're not real."

"I *am* real!" said Alice, and began to cry.

"You won't make yourself a bit realer by crying," Tweedledee remarked; "there's nothing to cry about."

"If I wasn't real," Alice said—half-laughing through her tears, it all seemed so ridiculous—"I shouldn't be able to cry."

"I hope you don't suppose those are *real* tears?" Tweedledum interrupted in a tone of great contempt.

Lewis Carroll
Through the Looking Glass

In the final analysis, the most recurrent and most distressing fallacy is the belief that regulation does not exist. In speaking about the fall of Rome, Eileen Power says:

All these people were deluded by the same error, the belief that Rome (the civilization of their age) was not a mere historical fact with a beginning and an end, but a condition of nature like the air they breathed and the earth they tread. *Ave Roma immortalis,* most magnificent most disastrous of creeds. (1963:15)

What human society has not been swindled by that belief—swindled while they themselves gobbled up their environment, or poisoned it, or severed their communications with it, or punished the very mechanisms that were trying to keep the society going for a few more years?

Even Eileen Power's words, written a few short years ago, have a taint of the same naïveté with which she labels the Romans. ". . . a condition of nature like the air they breathed . . ." Condition of nature, indeed! Like the air in New York, or London—or even smug San Francisco which used to laugh at Los Angeles smog. ". . . and the earth they tread . . ." Condition of nature, indeed! Like the rumblings of earthquakes in Denver, caused by pumping poison gasses into the ground "so they won't pollute the atmosphere."

When the ground rumbles, the people grumble—not before. Regulation is invisible—when it works. Once the government agrees to insure all bank deposits against runs, runs do not happen. But the regulatory system is actually working all the time—not so much by the insurance itself, but by the *publicity* keeping people informed that *if* there were a failure, they would be safe. To deny that there is regulation here just because little money is being paid out for bank failures is simple blindness to the true nature of the regulatory mechanism—the flow of information.

In government especially, regulation seems to be a target for abuse, but the accusations against "regulation" are almost inevitably misdirected. The agencies carrying the name "regulation" are hardly ever serving a regulatory function, because when an explicit mechanism is established—like the FCC or the ICC—it is soon coopted by the very industry it is supposed to regulate. But the truly powerful regulatory mechanisms go unnoticed as regulators.

For example, there are innumerable economic "stabilizers" built into United States law for ostensible purposes other than "stabilization." The income tax, for example, provides a number of regulatory mechanisms. In Figure 17.1, we see how income and spending tend to be in a reenforcing feedback loop regulated to some extent by proportional taxation, which removes more money from private spending when income is rising and less when income is falling. Or, since the tax scale is *graduated,* the taxes regulate the value of money, as sketched in Figure 17.2. If there are large government deficits, which tend to lower the value of the dollar, the effective tax rate will increase because the tax brackets are expressed in absolute amounts. Thus if everybody gets a double money wage—because the value of money is halved—many more people will be in "higher" brackets, even though their real wage is the same.

These examples are but the simplest of the "invisible" mechanisms which participate in the regulation of the economy. They, at least, are

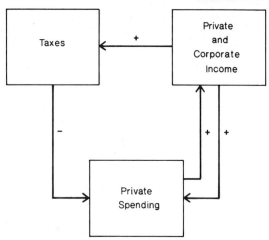

Figure 17.1.   How a proportional income tax tends to regulate spending and income through an "invisible" mechanism.

codified into law, but many more "invisible" mechanisms are simply built into the norms of behavior in the society. Such norms usually remain invisible unless and until specific revolutionary acts make them conspicuous, so that their existence may no longer be denied. Who knows how many of society's mechanisms incorporate the automobile until there is rationing of gasoline? Who imagines the influence of television on the birth rate until there is a power failure?

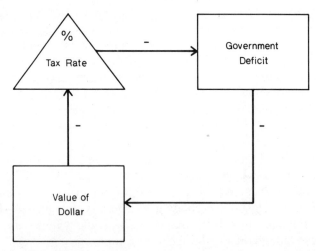

Figure 17.2.   How the graduated tax rates tend to regulate the value of money—another "invisible" mechanism.

The advent of space travel in our time has provided a number of revolutionary examples, such as the absence of gravity and atmosphere. The atmosphere provides us with a plethora of regulatory mechanisms whose existence was only hazily suspected until we thought of going beyond it. Of these, perhaps the most conspicuous is the regulation against meteorites. As one author wrote in *Science:*

. . . through the history of our planet, we have encountered this interplanetary debris, both large and small. . . . To assess their hazard to space travel we must have quantitative information concerning the total number of meteoroids in a given volume of space, the relative numbers of different sizes or masses, and the general nature of the motions around the sun.

And why do we need this information? To design an explicit regulatory system—a "meteor-bumper" or "meteor-shield"—for our spaceships. We don't need them on our automobiles, or on our heads, because the atmosphere does the work for us, silently, efficiently, relentlessly—and therefore, without recognition.

At times, of course, regulation is denied its role simply out of ignorance of its meaning when applied to a particular realm of thought. Consider the following quotation:

I consider it a dangerous misconception of mental hygiene to assume that what man needs in the first place is equilibrium or, as it is called in biology, "homeostasis," i.e., a tensionless state. What man actually needs is not a tensionless state but rather the striving and struggling for some goal worthy of him. What he needs is not the discharge of tension at any cost, but the call of a potential meaning waiting to be fulfilled by him. What man needs is not homeostasis but what I call "noodynamics," i.e., the spiritual dynamics in a polar field of tension where one pole is represented by a meaning to be fulfilled and the other pole by the man who must fulfill it. And one should not think that this holds true only for normal conditions; in neurotic individuals, it is even more valid. If architects want to strengthen a decrepit arch, they *increase* the load that is laid upon it, for thereby the parts are joined more firmly together. So, if therapists wish to foster their patients' mental health, they should not be afraid to increase that load through a reorientation toward the meaning of one's life. (Frankl 1963:166–167)

The truly "dangerous misconception" here is Frankl's. In the first place, regulation—homeostasis—is not an end in itself, but a *precondition* for the kind of activity Frankl describes. Were a person perpetually thinking about making his heart beat at the proper rate, he would hardly have time to answer the "call of a potential meaning." Were she occupied with the thousand-and-one other tasks which her body in its invisible wisdom does for her, she wouldn't even *hear* the call.

Frankl's analogy with physical structures is even more dangerous. Increasing the load on a structure only strengthens it when the structure is properly designed in the first place—that is, if all its parts are present and in the proper kind of relationship to meet this type of load. Certain varieties of mental illness may represent misstructuring rather than lack of load, in which case additional loading will surely hasten the patient's collapse. Or, given another kind of structure, the attempt at direct reorientation may flare back in the analyst's face. No doubt Frankl knows what he is *doing,* but what he is *saying* is pure nonsense. If psychiatrists really approached their patients with such crude actions, our mental hospitals would be more crowded than they are already.

But Frankl is not alone. Regulation is difficult to see, even for sophisticated investigators. The very nature of regulation makes it difficult to observe well regulated systems. For example, in biochemistry we find that:

High sensitivity of control systems is probably advantageous to the cell, but it adds to the difficulties faced by the metabolic chemist. Hypotheses . . . need to be tested in the living cell, and an obvious approach is to find whether concentrations of the metabolites involved vary in the expected ways in response to changes . . . But the more effective the regulation, the smaller the changes in concentration . . .

In other words, to study the structure of a system, you have to induce changes, but a well-regulated system resists such induction. To do experiments, we have to have "control," but control and stability are complementary. If the society is stable, there is little the government can do to control it—if by control we mean put it on some course that it is not following on its own. Indeed, the science of bureaucracy seems to consist in eradicating deep-seated regulatory mechanisms so that control may be exerted from the top. The same is true of scientific investigation—of society or cell—for as long as the regulatory mechanisms remain functional, they conspire against our knowing them.

Yet, though regulation may prevent us from probing its details, it does not prevent us from inferring its existence. We may not know *what* keeps us alive, but we may be sure that it is not simply our good looks or our direct communication with God. Eileen Power went on to say:

The fact is that the Romans were blinded to what was happening to them by the very perfection of the material culture which they had created. All around them was solidity and comfort, a material existence which was the very antithesis of barbarism. . . . How could they imagine that anything so solid might conceivably disappear? Their roads grew better as their statesmanship grew worse and central heating triumphed as civilization fell. (1963:15–16)

No regulatory mechanism works well forever, not even central heating. Environments change, and each change carries the seeds of disruption of a well-regulated existence. When regulatory mechanisms begin to fail— even before they begin to fail—new mechanisms must be constructed, or the system passes from existence. To remain the same in a changing world, everything must change.

QUESTIONS FOR FURTHER RESEARCH

1. *Microbiology*
Many bacteria have the ability to extract substances from their environment or to synthesize them. Discuss the problems and techniques of studying such bacteria in order to discover their synthesizing abilities.

2. *The Stock Market*
J. P. Morgan, when asked for a prognostication about the market, said, "It will fluctuate." If it did not fluctuate, would it be a market? Trace out the changes which might develop if the stock market suddenly stopped fluctuating. What institutions would be altered? Who would lose jobs and money? Which institutions would disappear altogether?

3. *Banking*
In what ways is deposit insurance advertised? What other mechanisms exist for ensuring the stability of banks?

4. *Cartels*
F. Scott Fitzgerald once wrote a story called "The Diamond as Big as the Ritz" about a family which lived on a mountain which was literally a diamond buried under a layer of soil. They made their living by chipping off small chunks from time to time, always careful never to reveal the source of their income or to sell too many diamonds at once. What would happen to the DeBeers diamond cartel if word got out about the diamond as big as the Ritz? What would happen if it were a gold nugget as big as the Ritz?

5. *Sociology*
What mechanisms exist to regulate the class structure of society— that is, mechanisms that tend to maintain the particular divisions and differentiation of people as they happen to be at any time?

6. *Space Travel*
In an earlier chapter, we discussed how gravity participates in the circulation of air essential to the breathing process. Try to list as many regulatory important mechanisms as you can in which gravity participates. Make notes of the difficulty you encounter in making this list, and try to

relate that difficulty to the ubiquity of gravity. Also, try to imagine mechanisms that could be constructed to replace gravity in these mechanisms, for the purpose of long space flights.

7. *Political Economy*

Richard Olney was an attorney for the Boston & Maine Railroad and Attorney General under Grover Cleveland. Once, when advising a railroad president, he said:

The (Interstate Commerce) Commission, as its functions have now been limited by the courts, is, or can be made, of great use to the railroads. It satisfies the public clamor for a government supervision of the railroads, at the same time that that supervision is almost entirely nominal.

Further, the older such a commission gets to be, the more inclined it will be found to take the business and railroad view of things. It thus becomes a sort of barrier between the railroad corporations and the people . . . (Wasserman 1972:29)

Do the railroads have to be this conscious of the evolution of regulatory agencies for that evolution to take place? Does the public have to be unconscious of the evolution for it to take place? What mechanisms could be established to retard, halt, or even reverse the process?

8. *Psychological Geology*

If you have ever been in an earthquake, describe your feelings the first time. Try to relate these feelings to the regulatory function of the earth beneath our feet, and how we take it for granted.

18

Epilogue

"It seems very pretty," she said when she had finished it, "but it's *rather* hard to understand!" (You see she didn't like to confess, even to herself, that she couldn't make it out at all). "Somehow it seems to fill my head with ideas—only I don't exactly know what they are!"

Lewis Carroll
Through the Looking Glass

Thus ends our study of regulation. Like the journey through the Looking-Glass World, it has been *rather* hard to understand—like that trick of ending the study of stability on an anticipatory note of change. Like Alice—like all children everywhere—we learn by filling our heads with contradictory ideas. Then we try to sort them out. For the most part, the study of jabberwocky is the best possible preparation for the study of the world of sensation. Anyone discomfited by nonsense will never make a mark as a systems thinker.

If we have succeeded, your head should now be filled with ideas. Don't be afraid or ashamed to confess that you "don't exactly know what they are" or even that you can't make it out at all. We *wrote* the silly book and can't claim to understand it.

If you're not even a little muddled, then obviously the book didn't do much for you, for if it's worth reading, it will be worth rereading. At least it has been for us. We've done our best. We hope it's good enough to do the same for you.

When this project began, 21 years ago, it seemed a simple enough task. It was to be a single volume, a simple introduction to a curious, general way of thinking. It expanded to two volumes. Then like a fertilized egg, it split into four. Now that two of the volumes are "finished," we see that it was really beyond the powers of our small group of people. And it always will be. The only thing that keeps us going is that we've also seen that it's beyond the powers of the other people who are writing on the subject, so it may be at least a stylistic contribution to the hesitant advance of knowledge.

Writing the third volume, on adaptive systems, seems no easier for having done the first two. On the contrary, it grows harder and harder the more we understand the problem. So we cannot say whether or not it will ever see the light of day. Certainly the generous response to the first volume has encouraged us in the work on this second one, so perhaps the same will be true for the third.

In case that wasn't clear, we're begging for compliments. But if not compliments, then *some* kind of response. If you have anything to say about it at all, we'd love to hear from you. After all, you've been listening to us for all these happy summer days.

QUESTION FOR FURTHER RESEARCH

There is but one question for further research in this chapter: How can the authors improve the clarity and usefulness of this volume, if there is a second edition?

SUGGESTIONS FOR FURTHER READING

Whether or not you were privileged to read them in childhood, or have them read to you, you should now quite definitely read Lewis Carroll's two *Alice* books. If possible, read them aloud to a 7-year-old child. If no child is available, an adult will do nicely.

A pleasant approach is to read first an unannotated edition, but with the original illustrations. Then obtain a copy of Martin Gardner's *Annotated Alice* and read it all again along with the notes. You won't be sorry.

Bibliography

Anderson, Robert T., and Gallatin Anderson
  1962   The Replicate Social Structure. Southwestern Journal of Anthropology 18:365–370.

Armer, Paul
  1963   Attitudes Toward Intelligent Machines. Datamation 9:34–38.

Ashby, W. Ross
  1952   Design for a Brain. New York: John Wiley & Sons.
  1956   Introduction to Cybernetics. New York: John Wiley & Sons.
  1964   The Set Theory of Mechanism and Homeostatis. General Systems IX: 83–97.
  1968   Principles of the Self-Organizing System. In Modern Systems Research for the Behavioral Scientist. Walter Buckley, ed. Pp. 108–118. Chicago: Aldine.

Babbage, Charles
  1826   On a Method of Expressing by Signs The Action of Machinery. Philosophical Transactions of the Royal Society, Vol. 2. (Reprinted In Charles Babbage and His Calculating Engines. Morrison, Philip and Emily, eds. New York: Dover, 1961.)

Barlow, R. E., and F. Proschan
  1965   Mathematical Theory of Reliability. New York: John Wiley & Sons.

Berge, C.
  1973   Graphs and Hypergraphs. Amsterdam, The Netherlands: North Holland.

Bernard, Claude
  1957   An Introduction to the Study of Experimental Medicine. New York: Dover.

Bloch, Marc
  1953   The Historian's Craft. New York: Vintage Books.

Blumer, Herbert
  1946   Collective Behavior. In Principles of Sociology. Alfred M. Lee, ed. Pp. 170–177. New York: Barnes and Noble.

Boissevain, Jeremy
  1974   Friends of Friends: Networks, Manipulators and Coalitions. New York: St. Martin's Press.
  1975   Introduction: Towards a Social Anthropology of Europe. In Beyond the Community: Social Process in Europe. Jeremy Boissevain and John Friedl, eds. The Hague: Department of Educational Science of the Netherlands.

Bosch, C. A.
    1971   Redwoods: A Population Model. Science 172:345–349.
Boulding, Kenneth
    1956   The Image. Ann Arbor: University of Michigan Press.
Bozon, Pierre
    1970   Le Pays des Villards en Maurienne. Grenoble: Allier.
Bridgman, P. W.
    1959   The Way Things Are. Cambridge: Harvard University Press.
Brillouin, L.
    1968   Thermodynamics and Information Theory: *In* Modern Systems Research
            for the Behavioral Scientist. Walter Buckley, ed. Pp. 161–165. Chicago:
            Aldine.
Bruner, J. S.
    1966   On the Conservation of Liquids. *In* Studies in Cognitive Growth. J. S.
            Bruner and R. Oliver, eds. New York: John Wiley & Sons.
Buckley, Walter, ed.
    1968   Modern Systems Research for the Behavioral Scientist. Chicago: Aldine.
Cabanac, Michel
    1971   Physiological Role of Pleasure. Science 173:1103–1107.
Cannon, Walter B.
    1939   The Wisdom of the Body. New York: Norton & Company.
Canon, Michael D., Clifton D. Cullum, and Elijah Polak
    1970   Theory of Optimal Control and Mathematical Programming. New York:
            McGraw-Hill.
Chargaff, Erwin
    1969   The Paradox of Biochemistry. Columbia University Forum 12(2):15–18.
Christie, Richard, and Florence L. Geis
    1970   Studies in Machiavellianism. Chicago: Academic Press.
Coale, J. J., and Paul Demeny
    1966   Regional Model Life Tables and Stable Populations. Princeton: Princeton
            University Press.
Coats, R. B., and A. Parkin
    1977   Computer Models in the Social Sciences. Cambridge, Mass.: Winthrop.
Coser, Lewis
    1956   The Functions of Social Conflict. New York: The Free Press.
Cruz, Jose B., Jr.
    1972   Feedback Systems. New York: McGraw-Hill.
Danch, William
    1971   How to Win Sweepstakes Prize Contests. New York: Fredrick Fell.
Dantzig, George B.
    1957   Concepts, Origins, and Use of Linear Programming. Proceedings, First
            International Conference on Operations Research. Baltimore: Operations
            Research Society of America.
Davenport, Horace W.
    1972   Why the Stomach Does Not Digest Itself. Scientific American 226(1):
            86–93.

DiStefano, J. J., A. R. Stubberud, and I. J. Williams
 1967 Block Diagram Algebra and Transfer Functions of Systems. *In* Feedback and Control Systems. J. J. DiStefano, A. R. Stubberud, and I. J. Williams, eds. New York: Schaum.

Durkheim, Emile
 1933 The Division of Labor in Society. George Simpson, trans. Glencoe, Illinois: Free Press. (First published 1893.)
 1951 Suicide. Glencoe, Illinois: Free Press Paperback. (First published 1897.)

Ehrenfeld, David W.
 1976 The Conservation of Non-Resources. American Scientist 64(Nov–Dec): 648–656.

Fisher, Ronald A.
 1958 The Genetical Theory of Natural Selection. New York: Dover. (First published 1929.)

Frankl, Viktor E.
 1963 Man's Search for Meaning. New York: Washington Square Press.

Fried, Morton H.
 1975 The Myth of Tribe. Natural History 84(4):12–20.

Goody, Jack, ed.
 1971 The Developmental Cycle in Domestic Groups. Cambridge: Cambridge University Press. (First published 1958.)

Goody, Jack and Ian Watt
 1963 The Consequences of Literacy. Comparative Studies in Society and History 3:304–345.

Gumbel, E. J.
 1958 The Statistics of Extremes. New York: Columbia University Press.

Gunning, Robert
 1952 The Techniques of Clear Writing. New York: McGraw-Hill.

Hale, Mason E., Jr.
 1970 The Biology of Lichens. New York: American Elsevier.

Hauser, P. M., and O. D. Duncan, eds.
 1959 The Study of Population. Chicago: University of Chicago Press.

Heer, David M.
 1968 Economic Development and the Fertility Transition. Daedalus 97(2): 447–462.

Hsu, Francis L. K.
 1972 Americans and Chinese. Garden City: Doubleday Natural History Press.

Hume, David
 1752 Political Discourses Of the Populousness of Ancient Nations. Edinburgh: Kincaid and Donaldson.

Hymes, Dell, ed.
 1969 Reinventing Anthropology. New York: Random House.

Jacob, F., and J. Monod
 1961 Genetic Regulatory Mechanisms in the Synthesis of Proteins. Journal of Molecular Biology 3:318–ff.

Keller, Helen
 1903 The Story of My Life. New York: Doubleday.

Keyfitz, H.
    1968   Introduction to the Mathematics of Population. New York: Addison-Wesley.
Kosambi, D. D.
    1966   Scientific Numismatics. Scientific American 214(2):102–111.
Krech, David, Richard S. Crutchfield, and Norman Livson
    1969   Elements of Psychology. New York: Alfred A. Knopf.
Kremyanskiy, V. I.
    1968   Certain Peculiarities of Organisms as a "System" From the Point of View of Physics, Cybernetics, and Biology. *In* Modern Systems Research for the Behavioral Scientist. Walter Buckley, ed. Pp. 76–80. Chicago: Aldine.
Kroeber, Theodora
    1961   Ishi in Two Worlds. Berkeley: University of California Press.
Kropotkin, Peter
    1970   Kropotkin's Revolutionary Pamphlets. New York: Dover.
Langer, Susanne
    1942   Philosophy in a New Key. Cambridge, Massachusetts: Harvard University Press.
Latané, Bibb, and John M. Darley
    1970   The Unresponsive Bystander: Why Doesn't He Help? New York: Appleton-Century-Crofts.
Lenin, Vladimir I.
    1929   What Is To Be Done? New York: International Publishers.
    1966   The Development of Capitalism in Russia. *In* Essential Works of Lenin. Henry M. Christman, ed. New York: Bantam Books.
Leontiff, Wassily W.
    1941   The Structure of the American Economy 1919–1929. Cambridge, Massachusetts: Harvard University Press.
Lin, S.
    1970   An Introduction to Error-Correcting Codes. Englewood Cliffs, New Jersey: Prentice-Hall.
Locke, John
    1947   Two Treatises of Government. New York: Hafner.
Lorenz, Konrad
    1964   Man Meets Dog. New York: Penguin. (First published 1953 in German. So Kam der Mensch auf den Hund.)
Lotka, Alfred J.
    1924   Elements of Mathematical Biology. New York: Dover. (First published 1924 as Elements of Physical Biology.)
MacAndrew, Craig, and Robert Edgerton
    1964   The Everyday Life of Institutionalized "Idiots." Human Organization 23(4):312–318.
Macaulay, Thomas B.
    1968   History of England. Vol. 4. New York: Dutton.
Macfarlane, R. G., ed.
    1970   The Haemostatic Mechanism in Man and Other Animals. Zoological Society of London Symposium, No. 27. New York: Academic Press.

Machiavelli, Niccolo
  1976   The Prince. Indianapolis: Bobbs-Merrill.

Malsom, Lucien
  1975   Wolf Children and the Problem of Human Nature. New York: Monthly Review Press.

Malthus, Thomas
  1964   Essay on Population. Ann Arbor, Michigan: University of Michigan Press. (First published 1798.)

Margalef, R.
  1963   On Certain Unifying Principles in Ecology. American Naturalist 97(897): 357–374.

Marshall, T. H.
  1929   Economic History. Vol. 1. London: Royal Economic Society.

Marx, Karl
  1957   The Eighteenth Brumaire of Louis Bonaparte. New York: International Publishers.

Maxwell, James C.
  1868   On Governors. (Reprinted *In* Modern Control Systems. R. Bellman, and Kalaba, eds. New York: Dover, 1964.)

May, R. M.
  1973   Stability and Complexity in Model Ecosystems. Princeton: Princeton University Press.

Mayr, Otto
  1970   The Origins of Feedback Control. Cambridge, Massachusetts: MIT Press.

McKim, Robert H.
  1972   Experiences in Visual Thinking. Monterey, California: Brooks/Cole Publishing.

Mills, C. Wright
  1967   The Sociological Imagination. New York: Oxford University Press.

Montessori, Maria
  1912   The Montessori Method. Anne E. George, trans. New York: Stokes. (Reprinted 1964. New York: Schocken Books.)

Moore, Barrington, Jr.
  1966   Social Origins of Dictatorship and Democracy. Boston: Beacon Press.

Morgenstern, Oskar
  1955   The Validity of International Gold Movement Statistics. Special Papers in Economics No. 2. International Finance Section. Princeton: Princeton University Press.

Nadel, S. F.
  1968   Social Control and Self-Regulation. *In* Modern Systems Research for the Behavioral Scientist. Walter Buckley, ed. Pp. 401–408. Chicago: Aldine.

Nett, Roger
  1968   Conformity-Deviation and the Social Control Concept. *In* Modern Systems Research for the Behavioral Scientist. Walter Buckley, ed. Pp. 409–414. Chicago: Aldine.

Nichols, J. R.
  1867   Chemistry of the Farm and the Sea. Boston: Williams.

Noble, Ben
    1969   Applied Linear Algebra. Englewood Cliffs, New Jersey: Prentice-Hall.
Oatley, C. W.
    1972   The Scanning Electron Microscope. Cambridge: Cambridge University
            Press.
Parsons, Talcott
    1964   Some Reflections on the Place of Force in Social Process. *In* Internal War,
            Problems, and Approaches. Harry Eckstein, ed. Pp. 33–70. New York:
            The Free Press of Glencoe.
Piaget, Jean
    1952   The Child's Conception of Number. London: Routledge and Paul.
Pierce, J. R.
    1961   Symbols, Signals, and Noise: The Nature and Process of Communication.
            New York: Harper & Row.
Piexto, M. M.
    1967   Qualitative Theories of Differential Equations and Structural Stability,
            Differential Equations and Dynamical Systems. New York: Academic Press.
Poincaré, Henri
    1952   Science and Method. Francis Maitland, trans. New York: Dover. (First
            published 1908 in French. Paris: Flammarion.)
Polanyi, Karl
    1957   The Great Transformation. Boston: Beacon Press.
Polya, George
    1945   How to Solve It. Princeton: Princeton University Press.
Power, Eileen
    1963   Medieval People. New York: Barnes and Noble. (First published 1924.)
Puck, Theodore T.
    1972   The Mammalian Cell as a Microorganism: Genetic and Biochemical
            Studies in Vitro. San Francisco: Holden-Day, Inc.
Rapoport, Anatol
    1968   Foreword. *In* Modern Systems Research for the Behavioral Scientist. Walter
            Buckley, ed. Pp. xiii–xv. Chicago: Aldine.
    1968   The Promise and Pitfalls of Information Theory. *In* Modern Systems Re-
            search for the Behavioral Scientist. Walter Buckley, ed. Pp. 137–142.
            Chicago: Aldine.
    1971   The Big Two: Soviet-American Perceptions of Foreign Policy. New York:
            Bobbs-Merrill.
Ritterbush, Phillip C.
    1968   The Art of Organic Forms. New York: Random House.
Rogers, A.
    1975   Introduction to Multiregional Demography. New York: John Wiley & Sons.
Russell, Bertrand
    1956   Portraits from Memory. New York: Simon & Schuster.
Sahlins, Marshall D.
    1976   The Use and Abuse of Biology. Ann Arbor: University of Michigan Press.

Savas, E. S.
　　　Mass Transfer and Urban Problems. Science. 174(4007):365.

Schneider, David M.
　　1968　American Kinship: A Cultural Account. Englewood Cliffs, New Jersey: Prentice-Hall.

Selye, Hans
　　1956　The Stress of Life. New York: McGraw-Hill.

Shannon, Claude E. and Warren Weaver
　　1949　The Mathematical Theory of Communication. Urbana: University of Illinois Press.

Sherrington, Charles S.
　　1906　The Integrative Action of the Nervous System. New York: Scribner.

Shock, N. W.
　　1957　Age Changes in Some Physiologic Processes. Geriatrics. 12(1)40–48.

Slater, Philip
　　1970　The Pursuit of Loneliness. Boston: Beacon.

Slicher Van Bath, Bernard H.
　　1968　Historical Demography and Social and Economic Development of the Netherlands. Daedalus. 97(2):604–621.

Spencer, Herbert
　　1904　The Principles of Sociology. New York: Appleton-Century-Crofts.

Steiner, Claude M.
　　1974　Scripts People Live. New York: Bantam Books.

Strehler, B. L.
　　1959　Origin and Comparison of the Effects of Time and High-Energy Radiations on Living Systems. Quarterly Review of Biology 34:117–ff.

Swets, J. A.
　　1973　The Relative Operating Characteristics in Psychology. Science. 182:990–1000.

Sworder, David
　　1966　Optimal Adaptive Control Systems. New York: Academic Press.

Taylor, Richard
　　1968　Purposeful and Non-Purposeful Behavior: A Rejoinder. *In* Modern Systems Research for the Behavioral Scientist. Walter Buckley, ed. Pp. 238–242. Chicago: Aldine.

Thistlethwaite, Frank
　　1964　Migration from Europe Overseas in the Nineteenth and Twentieth Centuries. *In* Population Movements in Modern European History. Herbert Moller, ed. Pp. 73–92. New York: Macmillan.

Thom, René
　　1975　Structural Stability and Morphogenesis. Reading, Massachusetts: Benjamin.

Thomas, Lewis
　　1974　The Lives of a Cell. New York: Viking Press.

Thrall, Robert M., and Leonard Tornheim
　　1957　Vector Spaces and Matrices. New York: Wiley.

Tomovič, R., and M. Vukobratovič
  1970   General Sensitivity Theory. *In* Modern Analytical and Computational Methods in Science and Mathematics. Rajko Tomovič, ed. New York: American Elsevier.

Tönnies, Ferdinand
  1955   Community and Association. London: Routledge and Kegan Paul. (First published 1887 in German.)

Vickers, Geoffrey
  1968   Is Adaptability Enough? *In* Modern Systems Research for the Behavioral Scientist. Walter Buckley, ed. Pp. 460–473. Chicago: Aldine.

Von Senden, Marius
  1960   Space and Sight. Peter Heath, trans. Glencoe, Illinois: Free Press.

Walker, Ernest P., and Associates
  1975   Mammals of the World. Baltimore: Johns Hopkins University Press.

Wasserman, Harvey
  1972   History of the United States. New York: Harper and Row.

Weinberg, Daniela
  1965   Models of Southern Kwakiutl Social Organization. General Systems. X:169–181.

Weinberg, Gerald M.
  1975   An Introduction To General Systems Thinking. New York: John Wiley & Sons.

White, Harrison C.
  1970   Chains of Opportunity: Systems Models of Mobility in Organizations. Cambridge, Massachusetts: Harvard University Press.

Whittaker, R. H., and P. P. Fenny
  1971   Allelochemics: Chemical Interactions Between Species. Science. 171:757–770.

Wiener, Norbert
  1948   Cybernetics. New York: Wiley.
  1954   The Human Use of Human Beings. Garden City: Doubleday Anchor.

Wilkins, Leslie T.
  1968   A Behavioral Theory of Drug Taking. *In* Modern Systems Research for the Behavioral Scientist. Walter Buckley, ed. Pp. 420–427. Chicago: Aldine.

Willens, Jan C.
  1971   The Analysis of Feedback Systems. Cambridge, Massachusetts: MIT Press.

Williams, J. D.
  1954   The Compleat Strategyst. New York: McGraw-Hill.

Wilson, Edward O.
  1975   Sociobiology. Cambridge, Massachusetts: Belknap Press of Harvard University Press.

Wolf, Eric R.
  1966   Peasants. Englewood Cliffs, New Jersey: Prentice-Hall.

Wrigley, E. A.
  1969   Population and History. New York: McGraw-Hill.

Wylie, Laurence
    1963   Demographic Change in Roussillon. *In* Mediterranean Countrymen. Julian
           Pitt-Rivers, ed. Paris: Mouton.
    1964   Village in the Vaucluse. New York: Harper and Row.
Zadeh, Lotfi A., and Charles A. Desoer
    1963   Linear System Theory: The State Space Approach. New York: McGraw-
           Hill.

Author Index

347

Subject Index